T0227843

A SIX SIGMA APPROACH TO SUSTAINABILITY

Continual Improvement
for Social Responsibility

Industrial Innovation Series

Series Editor
Adedeji B. Badiru
Air Force Institute of Technology (AFIT) – Dayton, Ohio

PUBLISHED TITLES

PUBLISHED TITLES

A SIX SIGMA APPROACH TO SUSTAINABILITY

Continual Improvement for Social Responsibility

Holly A. Duckworth

Andrea Hoffmeier

CRC Press
Taylor & Francis Group
Boca Raton London New York

CRC Press is an imprint of the
Taylor & Francis Group, an **informa** business

We would like to dedicate this book to our children. We hope that Manfred, Mathilda, Isabel, and Jonathan's children's children's children will be prospering in a sustainable world. We have hope that 1500 years from now we will have all evolved to understand the inextricable links between our people, our planet, and the opportunity to profit.

Contents

Preface

You have just picked up a toolbox full of tools. There are wrenches, screwdrivers, sockets, and saws. This toolbox will help you in your quest for improving the social responsibility performance of your organization. Continual Improvement for Social Responsibility (CISR®)* is a methodology for those interested in tackling the challenges currently facing every organization. We provide a way to *take action* to create a more sustainable environment where customers, supplier, employees, and communities are available for the organization for centuries to come. Yes, we mean centuries.

We take a cue from Kongo Gumi, a Japanese construction company launched in 578 AD and a thriving family-owned business until 2007 AD. That's almost 1500 years of continuous operation (O'Hara, 2004). Through more than a thousand years of emperors, wars, technology, and natural disasters, this company was sustained. Kong Gumi represents our goal of social responsibility. Our definition of sustainability means that your organization is successful for at least the next 1500 years. How will you work toward having a productive business environment with employees to add value, customers to purchase products and services, a safe and healthy community for

---

* CISR® (sounds like scissor) is a registered trademark and can be used with permission for non-commercial use. Contact SherpaBCorp.com for permission.

customers and employees, and a viable source of raw materials and suppliers 1500 years from now? What will your organization be doing in the year 3516 AD?

Applying business strategy and tactics with the intent of 1500 years of productivity is our vision for a social responsibility goal. This book documents and teaches a methodology for achieving this lofty goal. We borrow heavily from the scientific method, Six Sigma, and other product quality and process improvement methods and innovation tools. Because we are not teaching what to do, but rather how to approach perpetuating improvement, the methodology itself is sustainable. It is our goal that CISR is an applicable methodology in the year 3516 AD.

We have formulated and practiced this collection of tools focused on the improvement of social responsibility. It is an action-oriented methodology. It is a methodology that takes you step-by-step toward social responsibility performance improvement. The CISR toolbox is full of well-worn continual improvement tools borrowed from the profession of product and service quality improvement. These are not new tools; they are proven-successful tools used in a new way.

It is our hope that, after reading the following pages, you are motivated and able to take action and improve the social responsibility performance in your surroundings. We hope that we are preparing an army of zealots like ourselves. We hope that we are motivating a fervent group of others to preach from the mountain top. We hope that we are developing skills to be quickly applied. We hope that we change your attitudes and abilities and that you then change others. If we all touch just a few people and they touch a few people, then so on...you get the picture. Let's stop preaching and start doing.

Authors

Holly A. Duckworth is a certified Six Sigma Master Black Belt with over 30 years of experience in manufacturing. Holly leads organizations on the pursuit for process improvement. She has over 2000 hours as a classroom instructor. She is a volunteer leader for the American Society for Quality and their Social Responsibility Organization. Holly has coauthored two other books: *My Six Sigma Workbook* (Paton Professional Publishing) and *Social Responsibility: Failure Mode Effects and Analysis* (CRC Press).

Andrea Hoffmeier is a Six Sigma Black Belt with over 25 years' experience in marketing, product development, quality management, and organizational excellence. She has worked with organizations including Bellevue University, Callaway Golf, Convergys, Disney, First Data, Home Depot, Nordstrom, PS Brands, Ralph Lauren, Sunbeam, SunTrust Bank, UMUC, US Treasury, Verizon, Walmart and Xerox. She has trained and presented extensively on the topic of Social Responsibility, including articles in the American Society for Quality magazine and Wiley Journal. Andrea is completing her Master's degree in Sustainability and Environmental Management at Harvard.

Acknowledgments

We would like to acknowledge all of those who have touched our lives as we've become educated in social responsibility and sustainability. This includes Dave Celata at the American Society for Quality and Bob Pojasek at Harvard University. Roz Moore and Dorothy Bowers were also important teachers who helped us arrive at our SOFAIR method. Ellen Domb has provided invaluable insights and guidance over the years, on how to use TRIZ, the Theory of Inventive Problem Solving, to improve social responsibility performance.

Of course we couldn't have carved out the time to write without our husbands Gene Montano (for Holly) and Rob Wilson (for Andrea). We would like to thank that unseen Source of inspiration that planted the seeds of CISR and SOFAIR.

1

CONTINUAL IMPROVEMENT FOR SOCIAL RESPONSIBILITY

1.1 Introduction

Welcome to continual improvement for social responsibility (CISR®).[*] Let's orient ourselves to these, sometimes vague and nebulous, subjects of social responsibility and continual improvement. We take "continual improvement" from the tenets of the quality profession and product and process improvement. This methodology includes tools applied toward the goal of zero defect product quality and process capability, which minimizes waste and inefficiency. Organizational programs such as Six Sigma, Lean Production, Design for Six Sigma, Process Reengineering, and Total Quality Management are the foundations of our approach to continual improvement. The continual aspect addresses the need to never stop improving. Quality is an ideal state. The minute one level of performance is achieved, the customer expects the next level of performance. We can never rest. There is always a little more on which to chip away, hence *continual* improvement.

This improvement aspect is differentiated from change. It's easy to change. It's much harder to improve. Improving assumes that measurable progress to goals can be achieved and proven. Changing, for change sake, is not the goal. Measuring baselines, developing achievable and strategically important goals, identifying causation of performance levels, acting to improve or control input factors, and proving performance upgrades are all included in our definition of improvement. This level of performance improvement requires a technically rigorous methodology.

---

[*] CISR® (sounds like scissor) is a registered trademark and can be used with permission for non-commercial use. Contact SherpaBCorp.com for permission.

1

We take "social responsibility" from the tenets of ISO 26000:2010 (ISO, 2010). This is the International Organization for Standards guideline on social responsibility. As an internationally recognized document on social responsibility, it is applicable to global organizations. When using ISO 26000, there are seven core subjects: organizational governance, human rights, labor practices, the environment, fair operating practices, consumer issues, and community involvement and development. Social responsibility is not just environmental or human rights protection. We use these subjects to ensure that comprehensive targets are developed. However, the goal of social responsibility is the sustainment of productive suppliers, employees, customers, and communities (as described in the Preface, for at least 1500 years). And it is our contention that only when all seven of the core subjects are considered for responsibility that holistic sustainability, or social responsibility, is improved. In other words, social responsibility and sustainability isn't only about ecology or philanthropy. It is about performance in all seven subjects.

1.1.1 Our Purpose

Our purpose is to provide you with practical, actionable, and proven methods, which can be put to immediate use in any organization to improve social responsibility performance. We hope that you apply CISR as a way to provide a rigorous methodology for maximizing the efficiency and effectiveness of an organization's social responsibility performance improvement efforts.

Prior to focusing on social responsibility, our backgrounds were in leading the application of continual improvement tools for performance improvement in product and service quality. We have seen so many overlaps between social responsibility and continual improvement. We have seen many organizations either taking no action to improve their social responsibility performance or, worse, taking a marketing spin approach by broadcasting changes that are either false or are not relevant to their stakeholders. Our purpose is to provide you with a robust methodology to result in social responsibility performance improvement that delights your stakeholders.

We researched many organizations with very limited social responsibility human resources. In most large organizations, there are between two and five full-time professionals devoted to social responsibility or corporate social responsibility (CSR) (Duckworth, 2010). And yet, in these same organizations, there are large staffs devoted to continual improvement or product and service quality. With the addition of just a few tools, and a shift in focus, these existing resources could be deployed to improve social responsibility performance. Our purpose is to show you how to engage the resources that your organization already has in the continual improvement or quality assurance department, toward the application of social responsibility.

After reading this book, you should be able to take immediate action, adopt the tools we show you how to use, and equip the resources you already have. Achieving social responsibility performance improvement does not require the hiring of dozens of staff and it doesn't require the invention of a whole new tool set. Our purpose is to show you how to delight your stakeholders with the tools and resources already ready in your organization.

1.1.2 Our Audience

As we sat down to write this book, we had two primary audiences in mind. And, to be honest, it is our hope that, someday, these two audiences become one. First, we wrote this book for the quality practitioner. This includes quality managers, Six Sigma Black Belts, Green Belts, and Master Black Belts, and anyone else in the continual improvement field. As we have explained, quality is an ideal state and these professionals can never rest. Second, we wrote this book for the sustainability and social responsibility professional. This includes CSR leaders and executives, corporate affairs directors, sustainability officers, and social responsibility practitioners. Ultimately, we hope that the achievement of sustainability is embedded in the ideal of quality. Someday "quality" will mean excellent fit, form, and function for customers and sustainability for all stakeholders.

The quality practitioners will be very familiar with the tools and techniques presented in this book. They will easily understand the technical aspects of the continual improvement approach. These audience members may, however, struggle with the social responsibility core subjects, the stakeholder approach, and reporting aspects. However, educating this ready army of highly competent, potential social responsibility improvers may be the biggest boon to sustainability the world has ever seen!

The sustainability and social responsibility professionals will understand the need for stakeholder engagement and transparent and accountable reporting. They will easily absorb and navigate the myriad standards and guidelines. They will understand the need to approach all core subjects with equal measure. They may not yet appreciate the rigor of a methodology or the technical thoroughness of measuring and validating improvements. It is our hope that by teaching adapted but proven, continual improvement techniques to this audience of professionals we will significantly improve the integrity and validity of social responsibility initiatives.

But, really, our end goal is that the lines of these two audiences become blurred. We hope that everyone, in every organization, is prepared to lead continual improvement in social responsibility. We will have achieved our vision when it becomes expected, and even mundane, to have a 1500 year view of our decisions, actions, and impacts. When everyone in the organization behaves responsibly, when everyone in the organization strives for improving social responsibility performance, and when everyone in the organization lives sustainability, we will have deeply reached our intended audience.

1.1.3 How to Use This Book

As there are different audiences, there are different ways to use this book. Primarily, we aspire to motivate our readers to action. But it is ability, rather than motivation, that we target. The book is a "how to" guide. Recognizing that it takes both motivation and ability to mold behavior; we think motivation is best received when looking into a son's, daughter's, niece's, or nephew's eyes and recognizing that social responsibility improvement leaves our young ones in a better place to live. That's a stronger motivator than any book can deliver. Once the motivation is sparked, this book helps you put motivation to action.

One way to read this book is cover to cover. The first chapter helps to educate those unfamiliar with social responsibility and continual improvement. We start from scratch. We do not assume that our readers are familiar with either of these subjects. We also discuss a brief example of our Six Sigma approach to sustainability. Then, we discuss the stakeholders and subjects, objective, function and focus, analyze, innovate and improve, and report and repeat phases of the SOFAIR method. This is the technical detail of our method of performance improvement. Next, we provide many examples of SOFAIR being used in different settings. We have healthcare, manufacturing, business process, and personal life examples, all using SOFAIR for continual improvement of social responsibility. At the end of the book, we have a glossary, our references, and some additional reading for those readers who want to get even deeper into the subject.

In some organizations, there may be a lone, but motivated, supporter of sustainability. This person may see the benefits of social responsibility, but he or she is not quite sure how to start or how to convince others in the organization that social responsibility performance is possible. There are many organizations that don't embark on a formal social responsibility improvement initiative because they simply don't know where to start. The problem seems too big. We encourage these readers to share the examples in Chapter 5 with others in their organization. Start a grassroots effort. Show organizational leadership what is possible when a little bit of methodology is applied to a seemingly insurmountable problem.

Speaking of grassroots efforts, the spark for continual improvement for social responsibility often comes from people in the continual improvement effort of the organization. We encourage Six Sigma Black Belts and Green Belts to spend a lot of time with Chapter 4. See how we've adapted the Six Sigma toolkit for social responsibility. For these readers, the book is a partner to our practice of teaching people how to use SOFAIR. We keep busy by teaching the SOFAIR method to the problem solving and process improvement experts in these programs. Many of the techniques benefit from additional explanation and exploration. This book can be used as a companion to these classes.

We would like for this book to become a worn reference book. We would like for our readers to turn to our examples and methods any time they become stuck in a social responsibility quandary. Working through the phases of SOFAIR will help engage stakeholders, set objectives, focus effort, understand causal relationships, manage change, and report on improvement. Often by stepping through the method, centuries-old problems begin to unravel and solutions arise. Keeping this book handy as a reference ensures that the rigor of a methodology is deployed; with the validity of the methodology, "spin" and action with no positive outcome can be avoided. Only by following a rigorous methodology can we achieve lofty sustainability goals with integrity.

1.2 What Is Social Responsibility?

1.2.1 Sustainability as a Goal

Let's learn more about Kongo Gumi and the goal of sustainability. Kongo Gumi's primary business was building Buddhist temples (Hutcheson, 2007). Achieving successful temple construction for almost 1500 years required the successful performance of delivering a product or service to a customer, providing jobs to employees, having raw materials readily available and operating, being profitable to reinvest in the company, and living in an enjoyable place over centuries. In order for Kongo Gumi to succeed, they need a sustainable supply of wood. Buddhist temples are made of wood. Therefore, as a company with socially responsible behavior they would want to be good stewards of forests. They might sponsor the planting of new forests; they might ensure that they use varieties of wood that are easy to grow. Now, in order for Kongo Gumi to succeed, they need a sustainable supply of Buddhists! No Buddhists no temples. Therefore, as a company with socially responsible behavior they would want to sponsor the ongoing support of religious activities. They might donate to temples, provide scholarships for monks, and promote Buddhist values and philosophy. In order for Kong Gumi to succeed, they need a sustainable supply of workers. They need skilled builders and architects and designers. In modern times, they need civil engineers and materials specialists. Therefore, as a company with socially responsible behavior they might conduct outreach to elementary school children to motivate an interest in building professions. They might collaborate with trade schools and trade guilds to make sure that builders are well trained and capable of being good employees. In order for Kongo Gumi to succeed they needed to make a profit year after year. And they needed to take the profit and reinvest it in the company to modernize tools, increase capacities, and invest in technology. We use Kongo Gumi, the oldest company known to have existed, as an example of an organization achieving sustainability through social responsibility.

The end goal of social responsibility is sustainability. Our definition of sustainability is successful operations for at least 1500 years. Social responsibility is one way to achieve sustainability, but there are other ways to achieve sustainability. For example, many indigenous tribes have lived in the same place for thousands of years. They live in harmony with their local, natural environment. Their shelter, food, and entertainment are all provided from simple products, harvested in balance with what nature provides. If the tribe population exceeds what the local environment can produce, members starve and die, the population is reduced, and balance is returned. This is a tough but sustainable system.

Another way of achieving sustainability is through social responsibility. Social responsibility is defined as "the responsibility of an organization for the impacts of its decisions and activities on society and the environment, through transparent and ethical behavior" (ISO, 2010). Social responsibility involves both society and the environment. We must understand that, collectively, humans have the capacity to destroy natural environments. And we, as humans, only have the capacity to protect natural environments when we, as humans, have our society in order. No one cares about the environment in a war zone. Attention to social aspects is important to the environment.

The ISO 26000 definition stated above is important because it considers not just actions and decisions, but the impacts of those actions and decisions. We are responsible not only to our actions, but also what happens as a result of our actions. This definition is important because it tells us not just what to do, but how to do it…through transparent and ethical behavior. Achieving beneficial social and ecological actions and decisions without transparency and ethics is not socially responsible. Achieving sustainability through social responsibility requires us to carry out our decisions and actions with transparency and ethics and to be accountable to our impacts on both society and the environment.

Although we are teaching continual improvement for *social responsibility*, social responsibility is a means to an end. The end goal is sustainability. There are many possible paths to achieving sustainability. Indigenous life styles, sustainable development, and other approaches are valid paths to sustainability. This book focuses on one path to sustainability, social responsibility. And the path of social responsibility has some remarkable similarities to the quality movement of the mid- and late-twentieth century.

1.2.2 Quality Movement

In the 1950s and 1960s, W. Edwards Deming began teaching to businesses that would listen about the value of focusing on product quality and process control. Deming's "Fourteen Points for Management" (Deming's Fourteen Points for Management, n.d.), see Table 1.1, showed organizations how to *behave* to achieve a quality culture. Deming's "14 Points" was a shock to the manufacturing

Table 1.1   Deming's "Fourteen Points for Management"

1. Create constancy of purpose toward improvement of product and service, with the aim to become competitive and to stay in business, and to provide jobs.
2. Adopt the new philosophy. We are in a new economic age. Western management must awaken to the challenge, must learn their responsibilities, and take on leadership for change.
3. Cease dependence on inspection to achieve quality. Eliminate the need for inspection on a mass basis by building quality into the product in the first place.
4. End the practice of awarding business on the basis of price tag. Instead, minimize total cost. Move toward a single supplier for any one item, on a long-term relationship of loyalty and trust.
5. Improve constantly and forever the system of production and service, to improve quality and productivity, and thus constantly decrease costs.
6. Institute training on the job.
7. Institute leadership (see Point 12 and Chapter 8). The aim of supervision should be to help people and machines gadgets to do a better job. Supervision of management is in need of overhaul, as well as supervision of and production workers.
8. Drive out fear, so that everyone may work effectively for the company (see Chapter 3).
9. Break down barriers between departments. People in research, design, sales, and production must work as a team, to foresee problems of production and in use that may be encountered with the product or service.
10. Eliminate slogans, exhortations, and targets for the work force asking for zero defects and new levels of productivity. Such exhortations only create adversarial relationships, as the bulk of the causes of low quality and low productivity belong to the system and thus lie beyond the power of the work force.
    - Eliminate work standards (quotas) on the factory floor. Substitute leadership.
    - Eliminate management by objective. Eliminate management by numbers, numerical goals. Substitute leadership.
11. Remove barriers that rob the hourly worker of his right to pride of workmanship. The responsibility of supervisors must be changed from sheer numbers to quality.
12. Remove barriers that rob people in management and in engineering of their right to pride of workmanship. This means, inter alia, abolishment of the annual or merit rating and of management by objective (see Chapter 3).
13. Institute a vigorous program of education and self-improvement.
14. Put everybody in the company to work to accomplish the transformation. The transformation is everybody's job.

world when he started teaching them. They were one of the first attempts to connect culture to the outcome of a quality product. And Deming held management responsible for this outcome. Through his "14 Points" he was trying to teach managers how to behave. He recognized that leadership behavior made a difference in the tangible products delivered to the market.

Prior to Deming's influence (along with other early gurus of the quality movement), product quality was not an assurance. It was the time of "caveat emptor," let the buyer beware. Gradually, over decades, a philosophical shift occurred. First in manufacturing, and then with the delivery of service, quality assurance became a fundamental aspect of business management. Today, if a business expects to compete in the market, it must focus on product quality. The consumer today has very high expectations for product and service quality. And these expectations are constantly rising. This ever-increasing set of expectations, accompanied by ever-increasing performance, has created the profession of quality. We now have quality engineers, quality managers, and continuous improvement experts such as Six Sigma, Black Belts, and Green Belts. Each of these roles comes with bodies of knowledge and expertise in methodologies.

Let's better understand the quality movement with an example. Let's look at the healthcare industry. In the 1920s, the healthcare industry consisted of trained doctors and nurses. If you were sick, your doctor made a house call. If you were very sick, you would go to a hospital. At the hospital, there would be a few specialists such as nurses and surgeons. But typically you would see your personal physician. This person generally received university training. This training wasn't particularly standardized. Licensing was loose at best, especially in rural areas. The quality of healthcare you received was directly related to the skill of your family physician and where you lived.

In the 1950s, quality in the healthcare industry matured some. In many countries, by this time, there were standards for training and licensing. Nurses became professionalized with their own standards for training and licensing. There were hygiene standards in hospitals. And there were more specializations such as radiology and anesthesiology. Although the standards of training were improving, the quality of healthcare outcomes was not necessarily measured or a focus of improvement.

In the 1980s, significant research improvements were positively impacting the quality of healthcare. New drugs and new treatments were making a difference to outcomes. However, there was not a broad understanding that studying processes insured quality outcomes. There were still holes in the delivery of quality healthcare that led to the increase in another profession, the legal profession. Malpractice lawsuits were increasing as the consumers' expectations of quality increased. In some cases, the healthcare industry failed to rise to these expectations.

Now, in the 2010s, the healthcare industry has recognized the need to approach quality in service like many other industries. It is now common for major hospitals and providers to have statisticians, quality managers, and continuous improvement experts on staff. There is recognition that a formal study of process, process inputs and outputs, problem solving methodologies, statistical process control, and other quality assurance methods are necessary to meet the ever-increasing expectations of patients.

The maturity of the quality movement's impact on the healthcare industry can be seen and studied. 100 years ago, the relationship between doctor and patient was one of caveat emptor for the patient. Greater standardization and professionalization increased delivered quality to some extent, and increased expected quality a lot. The industry lagged the patient's expectations and legal tension ensued. Finally, a realization that quality methodologies need to be applied to stay ahead of patients' ever-increasing quality expectations has happened. We can see the same type of quality journey with other industries such as manufacturing, retail, finance, and education.

The quality movement is about culture change. It is about changing the beliefs, values, and behaviors of whole organizations, starting with management. As customers' expectations increase, product quality responds, thus further increasing customers' expectations. Soon, high levels of product or service quality are required to compete in the market. This leads to the adoption of philosophies like "constancy of purpose," "drive out fear," and "improve constantly" (Deming, n.d.). The quality profession is born and management practices are permanently changed.

1.2.3 Social Responsibility Movement

If we look closely, we can see similarities between the quality movement and a budding social responsibility movement. The same provider–stakeholder forces are at play. Ever-increasing expectations of stakeholders are driving changes in product and service delivery. And in most cases, the providers are lagging the stakeholder's expectations. In the social responsibility profession, there are very few standards, guidelines, gurus, boundaries, or tests of validity. There are very few credentials of professionalism. Training is inconsistent. And improvement methodologies are absent. In comparison to the quality movement, we are in the 1920s.

There are a few social responsibility guidelines; ISO 26000:2010 is the International Standards Organization's "Guidance on Social Responsibility" and there is the Global Reporting Initiative's G4 standard for sustainability reporting. These are solid foundations of guidance. There is SA8000 (SAI, 2014), a social accountability standard focused on worker and human rights that can be audited by a third party. There are a few seminal gurus such as Elkington (1998) and Lovins and Odum (2011). The European Organization for Quality has a Certified Social Responsibility Manager and Certified Social Responsibility Auditor program to credential professionals (EOQ, n.d.).

Performance improvement in most of these programs is not based on a consistent framework. Improvement is rarely addressed at a management system level. They are missing some topical gaps; we'll talk more about this later in the book. And they are all completely missing any sort of rigorous performance improvement methodology; that's why we are here to present CISR and SOFAIR. The field is new and just spreading its wings. Stakeholders are beginning to wake up, pay attention, and demand results. We believe this is the perfect place to start a movement. The quality movement did not gain steam from the producers. Demanding customers, who create a competitive advantage for quality products, were the fulcrum of the quality movement. The impetus for the social responsibility movement looks awfully similar.

1.2.4 Similarities to the Quality Movement

Now that the quality movement is mature, we see things like widely adopted management system standards. We have formal product and process improvement programs such as Total Quality Management, Six Sigma, Capability Maturity Model Integration, and Good Manufacturing Practices. We have learned from 100 years of wise gurus such as Deming, Crosby, Juran, and Feigenbaum. We have national quality awards and models of best-in-class performance. We have extended warranties and product liability laws. A professional industry exists to train and credential quality managers, quality engineers, auditors, and inspectors. We now know that not only is quality free (Juran, 1999), but quality also provides a competitive advantage that leads to profit.

It is our prediction that in a few short years, the same will be said of social responsibility. When the social responsibility movement achieves the same maturity as the quality movement, we will have global, auditable, management system standards. Most high-performing companies will have a social responsibility performance improvement program in place. Gurus will rise to the occasion to teach and lead in new thought and methods. Gradually, a profession and professional industry will emerge. There will be ubiquitous degree programs and professional certifications.

The good news for the social responsibility movement is that it will not have to start from scratch. If we're smart, we'll steal shamelessly from the quality movement. We will jump-start our standards, performance improvement programs, gurus, and credentials. So many similarities exist between quality and social responsibility we would be foolish not to borrow heavily from what has already been learned. There is a critically important similarity between quality and social responsibility. That is the idea of an ideal. We begin the quality journey in full recognition that perfect quality will never be achieved. We know that the potential, although perhaps small, always exists for a product failure or for customer dissatisfaction. Similarly, we also know that we will always have some negative environmental and social impact. Stakeholders will always demand just a little less impact, just a little more transparency and proactivity. The minute one level of performance is achieved, continual improvement to the next level of performance is immediately expected. "Quality" is an ideal. "Social responsibility" is an ideal.

This is different than an attitude that compliance is enough; this subtle difference makes a social responsibility initiative the perfect candidate for the application of many quality improvement tools. It is from the recognition of this similarity that this book has been conceived. We have recognized the nascent standards, concepts, measures, and trends. We have also recognized that many companies randomly stab at social responsibility performance improvement opportunities without a methodology to follow. We have taken the very well-worn, and wildly successful, tools of Six Sigma, and rather than applying it to product or service quality, we've adopted it to social responsibility. This took some changes. It took some trial and error. It isn't a one-for-one. But with a thorough understanding of social responsibility, we have been able to successfully innovate to create a worthwhile modification.

1.3 Current Trends in Social Responsibility

1.3.1 Business Imperatives

The field of social responsibility has been evolving for decades and the evolution is rapid, even as these pages are written. Thirty years ago, "the bottom line" referred exclusively to the profit or loss number on the last line of an income statement. In 1983, the United Nations tapped Gro Harlem Brundtland, former Prime Minister of Norway, to propose long-term environmental strategies for achieving sustainable development by the year 2000 and beyond. Despite the awareness that has been building for decades, our current state of global sustainability is arguably worse since the resulting 1987 Brundtland Commission report, which offers the definition of sustainability as "development that meets the needs of the current generation without compromising the ability of future generations to meet their needs" (Brundtland, 1987). In the 1990s, John Elkington (1998) coined the term "triple bottom line," referring to an organization's performance in social, environmental, and economic responsibility. This broader concept of value of wealth in society is reflected in the fact that "bottom line" has now become a metaphor for what people and organizations value most. For example, a person might declare, "Bottom line, I won't wear fur." An organization might publish the statement, "Bottom line, we do not use children to produce our products."

Twenty years since Elkington introduced "triple bottom line" to our lexicon, there has been an increasing focus on the need for the profitable business to be socially responsible, with all functions involved, from finance to operations and marketing. Ingrid Martin, Professor of Marketing at California State University, Long Beach explains, "The spirit of this philosophy has permeated the field of business through a marriage of the basic underlying paradigm that drives business—organizations must make a profit to survive in the long run—and the philosophy that we need to consider the impact of our consumption patterns on future generations" (Martin, 2015).

Today, many corporations report their financial, social, and environmental "triple bottom line" through formats such as the Global Reporting Initiative (GRI) (The Global Reporting Initiative, n.d.), or as part of their traditional annual reporting. The trend toward digital publishing of these reports, including speed of distribution and transparent access by the public, has increased the expectation that organizations share their successes and challenges with greater frequency than ever. An annual report is stale within months of publication and the public is now expecting a two-way dialogue, via social media, rather than merely receiving information. There are trends in reporting, including the impact of social media on the volume and pace of communication.

Evidence from the hundreds of thousands of daily social media communication threads is clear. Organizations across the globe are becoming increasingly aware of the need for, and benefits of, operating in a socially responsible way. Increasing competitive advantage and elevating reputation are just a few reasons for focusing on social responsibility. And through social media, customers and investors have a more influential voice. This awareness has led to some top-down approaches to improving social responsibility performance

In 2004, the first Chief Sustainability Officer (CSO) was appointed in the United States by Dupont (Bader, 2015). However, Bader asks, "How can onlookers know whether a CSO (or equivalent by another name) is driving real change, or is just the Chief Greenwasher?" Greenwash refers to misleading information about a company's environmental impact or efforts. In other words, is the appointment of a CSO merely window dressing? Or is the CSO legitimately changing culture and improving performance?

While sustainability has become associated with environmental or "green" issues, we take a more holistic view. We expand our concern to overall "social responsibility spin," which is disingenuous, or misleading marketing and public relations about an organization's social responsibility performance. Bader (2015) observes that most, if not all, of the companies involved in every major recent corporate disaster, such as the 2010 Deepwater Horizon oil spill and the 2013 Rana Plaza factory collapse, had executives with such (CSO) titles. This observation leads to the notion that a CSO might not only be ineffective, but also even be a harbinger of bad behavior. Other thought leaders in the social responsibility field have objected to the position of CSO, in that social responsibility should not be a "bolt on" function, or figure head, but rather embedded in behavior and methods throughout the organization. And we would add, how could the CSO affect change without a performance improvement methodology?

While it is encouraging to see many more companies create a senior executive position responsible for sustainability or CSR, our objective is to embed social responsibility into organizations systemically. CISR practitioners have the potential to transform the world by merging the existing quality discipline with social responsibility.

CISR is a new opportunity and a new framework for quality professionals to bring discipline and valuable expertise to social responsibility efforts. Conversely, other professionals involved in improving on, and reporting about, social responsibility performance, now have well-worn tools and a process that they can use with confidence. CISR can be used in any type or size of organization, ensuring that social responsibility efforts yield meaningful results, rather than just glossy marketing spin.

The perception and reality of your organization's social responsibility performance can influence your ability to attract and retain employees, customers, and investors. The more socially responsible your organization is, the more welcome you will be in the communities where you operate and the better you will attract top talent (Santos, 2013). In the next section, we will examine ISO 26000, the International Standard for Social Responsibility, and its seven subjects. And we will find that CISR can be used not only for risk mitigation, but also for competitive advantage.

The publication of ISO 26000 in 2010 was an important milestone in galvanizing the current business imperative for social responsibility.

Prior to that landmark achievement, there were differing views globally on what it means to be socially responsible. The challenge with ISO 26000 is that it is a standard that was published as a guideline and not a certifiable standard via third-party auditing. With other familiar ISO standards, such as 9000 for quality, 14000 for environmental issues, or 31000 for risk management, there is a clear path to proving compliance via external auditors. Certification in these other standards is an expectation in many industries. However, ISO 26000 was published as guidance only. At this time, your organization cannot be certified to ISO 26000.

The ISO 26000 Working Group on Social Responsibility, the writers of ISO 26000, consisted of about 500 delegates from across the globe, from industry, government, nongovernmental organizations (NGOs), labor and consumer organizations, and a broad group called SSRO—service, support, research, and others (primarily academics and consultants). These delegates met over five years, from its first plenary session in 2005 to the final official vote in September 2010, with the standard published November 1, 2010. This ISO standard is now seen as the seminal resource for defining what social responsibility is, what it includes, and how to integrate social responsibility in an organization. We will use ISO 26000 as our definition and guidance on social responsibility.

1.3.2 ISO 26000 Core Subjects

By defining the core subjects of social responsibility, ISO 26000 can usher in a new megatrend for social responsibility, the same way ISO 9000 did for quality in the late 1980s. Understanding these seven core subjects, thoroughly, will make the difference between superficial involvement versus establishing an influential role in social responsibility efforts. In ISO 26000, there are nearly 40 issues addressed across the seven subjects. We will discuss these subjects and issues in detail. With our approach, in order to focus on legitimate social responsibility performance improvement, improvement in these subjects will need to be proven through the CISR methodology. Now, we will explain these cores subjects in detail. The seven subjects are: organizational governance, human rights, labor practices, environment, fair operating practices, consumer issues, community involvement, and development. It is important to understand all seven subjects in order to understand the breadth of a topic like social responsibility.

1.3.3 Organizational Governance

Organizational governance refers to the way a business is run. Rather than describing issues of organizational governance, ISO 26000 describes seven principles associated with this subject. The principles are accountability (making sure that people who act for the business are held responsible for their actions), transparency (openness in explaining how the business operates, makes decisions, handles money, etc.), ethical conduct (treating others with honesty and fairness), consideration of stakeholders' interests, and obeying the laws. Following good governance principles will help the business to improve its social, environmental, and economic triple bottom line. This benefits the business, all of its stakeholders, the natural environment, and the community in which the business operates. For each subject, we will list activities that demonstrate the intent of the subject. Here are the action suggestions for organizational governance.

Organizational governance action suggestions:

- Create and model a company culture where these seven principles are practiced.
- Commit to respecting laws, including the responsibility to pay taxes to the government bodies and communities in which you operate.
- Efficiently use financial, natural, and human resources, while ensuring fair representation of historically underrepresented groups (including women and racial and ethnic groups) in senior positions in the organization.
- Balance the needs of the business and its stakeholders, including immediate needs, and those of future generation. Take a 1500 year viewpoint of organizational strategy.
- Establish a permanent two-way communication process between the organization and stakeholders. This has never been more doable than in today's world. Think about social media and online communities.

- Encourage greater participation by employees in decision making on social responsibility.
- Delegate authority proportionately to the responsibilities assumed by each member or employee in the organization.
- Keep track of decisions to ensure they are followed through, and to determine responsibilities for the results of the organization's activities, either positive or negative.
- Leadership is critical to effective organizational governance, which should be based on incorporating the seven principles of social responsibility into decision making and actions.
- Conduct external auditing to ensure regulatory requirements are being met.

1.3.4 Human Rights

The subject of human rights refers to respectful treatment of all individuals, regardless of any of their personal characteristics, just because they are human beings. There are eight issues to consider: due diligence, risk situations, avoidance of complicity, discrimination, civil and political rights, resolving grievances, economic, social, and cultural rights, and rights at work. These issues might include vendor selection practices which do not consider countries of origin known for violations of political rights, employment practices which inadvertently promote human trafficking, or discriminating against the disabled. To be socially responsible, an organization should try to increase its "sphere of influence" on human rights. Here are some action suggestions on human rights:

- Create vendor selection practices especially when considering vendors in developing nations that assure respect for human rights.
- Avoid complicity by making your best efforts to find out how people are treated by other organizations in your supply chain and value chain. Take measures to prevent cruel, inhuman, or degrading treatment, and the use of excessive force by those you do business with, or else stop doing business with them.

- Test for discrimination and vulnerable groups. Monitor employment statistics for these groups and ensure that the organization's employment does not demonstrate discriminatory practices.
- Participate in transparent and accountable political processes in areas where the organization operates. Ensure that stakeholders have a voice in the political process.
- Contribute to the protection of the rights of indigenous people to sustain an indigenous lifestyle.

1.3.5 *Labor Practices*

Labor practices refer to fair treatment of all workers, including those who are full employees and those who are subcontracted. Labor practices include hiring and promotion of workers; disciplinary and grievance procedures; transfer and relocation of workers; termination of employment; training and skills development; health, safety and industrial hygiene, and any policy or practice affecting conditions at work. Labor practices also include recognition of worker organizations and participation in collective bargaining, including tripartite (business–workers–government) consultation to address social issues related to employment. ISO 26000 guides that human labor is not a commodity; because workers are human beings, they need protection, and their treatment should not be governed by the same market forces that apply to commodities. The following are action suggestions for labor practices:

- Recognize employees as key stakeholders. This is inherent in the employment relationship.
- Ensure safe and healthy working conditions.
- Develop means of social dialogue with employees. It is not necessary, in most places, to accomplish this only through collective bargaining units. If a system of stakeholder engagement is functioning, and employees are treated as stakeholders, social dialogue will ensue.
- Work with the community and external organizational environment to ensure training and human development resources are available to current and future employees.

1.3.6 Environment

There are four issues identified by ISO 2600 with respect to environmental sustainability: prevention of pollution, sustainable resource use, climate change mitigation and adaptation, and protection and restoration of the natural environment. Sustainable resource use should be planned throughout the life cycle of a product. A failure to recognize environmental impacts for the full life cycle of products and services is the risk for this issue of social responsibility. Sustaining an abundant ecology for future generations is the goal. A minimization of resource waste and the reduction of impacts will improve the issue of the environment as an element of social responsibility. Following are some action suggestions for the environment:

- Measure all outputs of the organization that impact the environment. Use continual improvement efforts to ensure the perpetual improvement of this impact.
- Ensure that sustainable resource is a key factor in product design. This should include the recyclability of products at the end of the product life.
- Risks due to climate change should be embedded in business strategy and planning. This is accomplished for the protection of owner stakeholders. Investing in coastal areas which may be significantly affected by climate change may be a risk to owners.
- Use organizational resources to repair wild habitat where appropriate.

1.3.7 Fair Operating Practices

Fair operating practices are concerned with building systems of fair competition. The issues include preventing corruption, encouraging fair competition, and promoting the reliability of fair business practices helps to build sustainable social systems. This subject is interested in how organizations treat each other. An organization which is involved in the political process to ensure fair and equitable laws, taxes, or oversight is helping to build a sustainable society.

An intention to operate unfairly prevents the development and maintenance trust in a fair market. Here are some action suggestions for fair operating practices.

- Develop auditing systems to prevent bribery at all levels of the organization. Seek to prevent, not correct, corruption in organizational activities.
- Use transparent and accountable practices with the organization's political involvement. Share political action investment decisions with stakeholders.
- Promote fair market competition in operating regions.
- Carefully consider expansion impact. Organizational growth should not come at the expense of community member property rights. Consider expanding operations in regions where community benefit can be reaped from the expansion.

1.3.8 Consumer Issues

Consumer issues are primarily concerned with product safety and quality. This subject is concerned with the protection of the rights of the consumer. Protecting the health and welfare of consumers helps to build trust and accountability to stakeholders. There are seven issues: fair marketing, information, and contracts, protection of consumer safety and health, sustainable consumption, consumer support, consumer data protection and privacy, access to essential services, and education and awareness. Honest and thorough communication between the provider and the consumer is responsibility. This includes a concern for the consumer experience, including the information shared in that experience. Protecting the consumer from adverse impact caused by a product or service is socially responsible. Here are some action suggestions for consumer issues:

- Provide easy to read information (not in legal-ese or engineer-ese) to consumers about contractual details on products or services.
- Ensure product safety. Use third-party test laboratories for verification of product safety.
- Ensure sustainable consumption. Market solutions. Do not design excess into the product offering.
- Offer service after the sale. Ensure that consumers have easy access to product or service issue resolution.
- Design consumer information privacy into the service design.

1.3.9 Community Involvement and Development

The core subject of community involvement and development intends to create sustainable environments where increasing levels of education and well-being can exist. Actionable social sustainability starts in the local community. There are seven core issues of community involvement and development: community involvement, social investment, employment creation, technology development, wealth and income, education and culture, and health. A focus on the needs and benefits of the local community is socially responsible. Therefore, social sustainability begins with community development and involvement, and community involvement is the means by which social responsibility is deployed. Here are some action suggestions for community involvement and development.

- Encourage employees (especially those in leadership roles) to be involved in community organizations. Allow for time off from work for this participation.
- Understand the cultures of the local people. For global organizations, the goal is to appreciate and sustain local culture, not to spread the organization's home culture to other areas.
- Expand into new communities with a 1500 year perspective. Plan how the organization will be interacting with the community over very long periods of time. Have a multigenerational approach to community development.

2

THE STORY OF CONTINUAL IMPROVEMENT

Improving products, processes, and services to make them better for customers is a very old practice. Let's go back to the Kongo Gumi example and its origins in 578 AD. At that time, in Japan, businesses were handed from father to son. Sustaining a business was predicated on the birth of boys. However, in the 1930s, Kongo Gumi, a very successful construction company, found itself without male heirs (O'Hara, 2004). Rather than closing the company, they chose to be innovative. It is reported that Kong Gumi used the very unconventional practice of adopting the husbands of daughters as sons, and thus, sustain the line of heirs and therefore the continuity of the company. The point is that sustainable companies improve and innovate. Sustainability isn't about stasis; it is about continual improvement. As attributed to W. Edwards Deming, "Survival is optional. No one has to change."

In the following sections, we will provide information to better understand the history of important performance improvement and innovation methodologies. As we stand on the brink of the beginning of a formal social responsibility performance improvement methodology, CISR®,* it is important to learn from our predecessors. We will also discuss continual improvement programs and their role and tactics in organizations. We advise continual improvement for social responsibility to be integrated into existing continual improvement programs; therefore, understanding how these programs work is important to our ability to integrate CISR into the organization.

---

* CISR® (sounds like scissor) is a registered trademark and can be used with permission for non-commercial use. Contact SherpaBCorp.com for permission.

2.1 The History of the Quality Movement

We've discussed the nature of the Quality movement and how it relates to social responsibility. Now let's dive a little deeper into the history of the Quality movement. Perhaps if we understand how we got to where we are with Quality, we can better understand where we are going with social responsibility. The story of continual improvement takes us back to Bell Labs; back to when Alexander Graham was still alive.

In 1918, Walter Shewhart was a newly graduated physicist when he joined Western Electric (Shewhart, 1931). His early work focused on identifying the difference between "chance-caused" variation and "assignable-cause" variation using statistics. Shewhart noted that reducing variation in the manufacturing process was critically important, but too many adjustments to the process, especially when shifts and trends were due to chance-caused variation, increased variation. Studying variation before adjusting the process is required. Unstudied response and adjustments increase variation. It is upon this simple premise that the entire field of continual improvement is built.

As Shewhart was adding to the body of knowledge of quality control, mass production was taking off. This was the heyday of Henry Ford, the birth of assembly lines and industrial process engineering. At the end of most of these assembly lines, there were large organizations of inspectors. The modus operandi was to wait until the end of a given process, physically inspect the product delivered, note defects, and then report patterns and trends back to the originating process to attempt to correct or prevent the root cause of the defects. The delay in information between inspector and producer could be minutes, weeks, hours, or even months. In the meantime, a lot of scrap and cost was incurred.

The next big leap in the Quality movement came from the influence of W. Edwards Deming. Deming, like Shewhart, was degreed as a physicist and worked for the U.S. Census Bureau (ASQ, n.d.). He perpetuated the use of Shewhart's work to nonmanufacturing problems. However, Deming's big opportunity was created by war. He worked with the War Department during World War II, and was sent to Japan in 1946 by the Economic and Scientific Section of the War Department to help in the effort to rebuild war-ravaged Japan. Prior to the war, Japanese manufacturing output was notorious for poor quality. As a consultant to many Japanese companies rebuilding after the war he demonstrated, through his management principles, that organizational culture could be reversed. Using Shewhart's statistics and his own brand of no-nonsense management strategy, such as his statement on Profound Knowledge (Table 2.1), Deming famously sparked the Japanese quality revolution, from which the rest of the world was to later study.

It is important to note that Deming was a proponent of Shewhart's "Plan–Do–Study–Act" (PDSA) methodology. This may be the first process improvement methodology of the modern era. Shewhart was frustrated with the over adjustment, and thus variation inducing, practices in manufacturing. The "Plan" and "Do" phases of his methodology required careful and intentional experimentation. Only after this "Study," or cause and effect analysis, of the results of careful experimentation should the process be adjusted or modified. Deming made this PDSA methodology a foundational element of his work; perhaps inventing the concept of continual improvement.

The next leap in the Quality movement can be attributed to A. V. "Val" Feigenbaum. In the 1970s, Feigenbaum was introducing his concept of "Total Quality Control," later known as Total Quality Management or TQM (1961). With this concept, Feigenbaum set the expectation that a system for integrating the quality development, quality maintenance, and quality improvement efforts in order to provide customers with the best quality and at the lowest cost. He espoused management accountability to quality-related performance metrics. Feigenbaum, educated as an economist, brought Quality out of the laboratory and off of the shop floor, and into the boardroom.

This brings us to the modern era of the Quality movement. Quality assurance is a credentialed profession; the American Society for Quality (ASQ), a professional society of those in the Quality field, has more than 80,000 individual members as well as enterprise

Table 2.1 Deming's System of Profound Knowledge

---

1. *Appreciation of a system:* Understanding the overall processes involving inputs, outputs, transformations, and feedback.
2. *Knowledge of variation:* Understand the range and causes of variation in the system.
3. *Theory of knowledge:* Understanding what is known and what is unknown, and what is unknowable.
4. *Knowledge of psychology:* Understand the interactions of the human and the system.

---

members consisting of many large global companies. ASQ provides certification for 18 different professional levels (ASQ, n.d.). The European Organization for Quality provides certification for 49 different professions, including social responsibility manager and social responsibility auditor (EOQ, 2015).

One of the quality professions that has experienced the fastest growth in the twenty-first century is that related to Six Sigma: Green Belt, Black Belt, and Master Black Belt. Six Sigma is the pinnacle of the current state of process improvement. Standing on the shoulder of Shewhart's statistical methods, Deming's culture change model, and Feigenbaum's boardroom attention, is Six Sigma.

2.2 The History of Six Sigma

In the 1980s, Motorola was faced with severe competition with their mobile phone devices. Nokia was out-competing; Motorola was losing significant market share due to product quality issues. Motorola's historical measurement of defects in thousands of opportunities didn't motivate the kind of problem solving that was needed to compete. They moved to measure the defects per million opportunities. Six Sigma is the term used to describe product performance that is six standard deviations (the mathematical term for standard deviation is "sigma") better than the customer's expectations. This level of performance can also be described as 3.4 million defects per opportunity. Six Sigma was begun as a stretch goal of product quality at Motorola in order to regain lost market share.

In order to achieve these near-perfect levels of performance, a new problem-solving methodology was needed. This sparked, at Motorola, the define–measure–analyze–improve–control (DMAIC) problem-solving method. DMAIC is an innovation to the scientific method. We will discuss it in a lot more detail, later in this chapter. DMAIC ensures an extreme level of rigor, statistical proof, and carefully controlled process change in consideration of delighting customers. A significant innovation of Six Sigma is the concept of full time performance improvement experts, called Black Belts. Allocating resources to the pursuit of performance improvement makes a difference.

Most large companies around the world have adopted Six Sigma as a way of doing business. Six Sigma has evolved over time, to include service quality, healthcare quality, and quality in design. It continues to be an important differentiator of companies that compete, and win, with product and service quality. Six Sigma became a way to translate classical problem solving and performance improvement in a way that business leaders could understand and the imperative of resources this type of work demands.

2.3　History of the Theory of Inventive Problem Solving

We wanted to include a little bit about the theory of inventive problem solving (TRIZ) in this chapter. Understanding what TRIZ is, how it is used, and how it came about will be important later on. TRIZ, pronounced "tree-z," is an acronym for the Russian words meaning "theory of inventive problem solving." It is an innovation methodology. Since we see CISR as a disruptive innovation to continual improvement methods, it's important that our innovative methods of improvement contain methods of innovation.

The development of TRIZ is attributed to Genrich Altshuller, a Russian scientist, who became well known in Russia in the 1970s. Altshuller was a clerk in a patent office (like Albert Einstein) and he began to study patterns of solutions from the patents (Altshuller, 1996). He began to notice some commonalities across solutions and then created algorithms for solving any problem using matrices of comparison between the problem and many possible solutions. The TRIZ methodology is a process of convergent and divergent thinking that takes the innovator through a process of modeling the problem as a set of contradictions, separates the aspects of the problem using a set of principles, and then tries substitution, combination, alternatives, and other approaches to find solutions to the problem (see Figure 2.1). Altshuller created a systemic methodology for innovation.

TRIZ is taught and used as a tool in the Six Sigma toolbox, particularly in the subfield of "Design for Six Sigma" (DFSS). Often, when tackling a particular difficulty, continual improvement challenge innovation is needed. And in consideration of delivering Six Sigma quality (performance that meets or exceeds the customer's expectations by at least six standard deviations), robust designs are needed. TRIZ is a proven innovation technique used to deliver this level of design performance. Intel, Ford, Boeing, and NASA, have reported

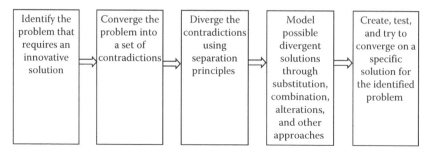

Figure 2.1 The process of TRIZ.

success by using TRIZ for more notable innovations (TRIZ-Canada Organisation, 2015).

We will be using TRIZ in the CISR toolbox, too. There are many of our social responsibility problems that are chronic, multipronged, and difficult to solve. We will need innovative solutions. CISR practitioners will need rigorous innovation methodologies, like TRIZ, to deliver socially robust solutions to our toughest sustainability problems. In later chapters, we will go in detail on how to use TRIZ as a part of CISR.

2.4 Why Organizational Structure Is Important

As we've discussed previously, continual improvement is done to achieve an ideal state, be the topic quality, productivity, or social responsibility. And yet the ideal state is never really achieved because the customers' or stakeholders' expectations are ever-rising. For these perpetuating reasons, continual improvement becomes an integral part of the organization supported by resources, structure, authority, and budget. The following section is devoted to understanding the successful organizational structure of continual improvement programs. Later, we will discuss how this structure can be translated to include continual improvement for social responsibility.

In many cases, continual improvement programs are launched in order to affect the culture of the organization. Our customers are dissatisfied, or we have failed to gain new business, or our market share is decreasing due to innovation by our competitors. These might be externally motivated "burning platforms" on which an organizational culture change is launched. In other cases, we might have new ownership or leadership who has seen continual improvement in action and understands the opportunities for performance improvement that are being missed. Or we might have high employee attrition, or employee health and safety issues. There might be internally motivated reasons to launch an organizational culture change.

In any case, culture change does not happen easily or quickly. Changing organizational culture means changing the behavior of everyone in the organization, permanently and all of the time. A critical mass of resources focused on the culture change is needed. And these "change agents" must have a rigorous methodology upon which everyone in the organization can focus in order to spark the culture change. To put it in mechanical terms, we need a fulcrum (the methodology) and a lever (the change agents) in order to move the heavy weight (organization) to a new place (culture change). See Figure 2.2 for a graphical depiction. The implementation of these resources, the fulcrum and the lever, are typically deployed as a focused department within the organization. Often, the weight is so heavy, and the new place so far away, that intense focus and effort will be needed. Changing the culture of the organization isn't something you do in addition to your "day job"; it is a full time job.

The organizational structure of the continual improvement effort is important when achieving a culture change. And having a separate continual improvement (CI) department offers many benefits. The CI group can garner top leadership visibility and management support. Resources can be ramped up or down based upon the maturation of the culture change. Extra resources can be deployed to surge an area of the company that might be struggling with the culture change. The CI organization can be held accountable for metrics and measure of the culture change. And methodology experts can be groomed and developed within the CI organization.

However, it is important to remember that this CI organizational structure is only needed *during* the culture change. The CI

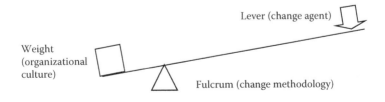

Figure 2.2   Lifting the weight of culture change.

organization is the lever to move the heavy weight. Once the weight is moved, the lever isn't needed. It may take 5, 10, or 20 years, but ultimately the end goal is to not have a separate CI organization. The end goal, for the CI department, is the culture change. The end goal is that everyone in the organization is always striving to achieve internal and external peak performance. Once this is achieved the organizational structure changes again.

2.5 Important Factors in a Continual Improvement Program

Not all CI programs are made alike. With our combined 40+ years of leading and operating with CI programs we've seen successes and failures. There are a few nonnegotiable factors that make the success of the CI program more robust (with success of the CI program measured by the culture change). Organizational support, clear roles and responsibilities, a rigorous methodology, linkage to the business strategy, and measurable outcomes are critical factors in the success of the CI program (see Table 2.2). Later in this book, we will translate these success factors toward continual improvement for social responsibility. In most organizations, the CI program is focused on operational performance improvement such as safety, quality, cost, or delivery performance. For the moment, let's focus on the factors for a general organizational culture change, then later we will focus on a social responsibility culture change.

Organizational support, as a critical success factor in a CI program typically starts with top leadership's visibly communicated mandate for the culture change. Top leadership should communicate the compelling reason for the need of the culture change. This message is best delivered in three parts: What terrible thing happens if we fail to change, what wonderful thing happens after we change, and the path from terrible to wonderful. The message needs to be linked to

Table 2.2 Critical Success Factors of a Continual Improvement Program

1. Organizational support
2. Clarity of roles and responsibilities
3. Rigorous methodology
4. Linkage to business strategy
5. Measurable outcomes

real business issues or opportunities. All members of the organization need to be able to see, feel, and hear the crackling of the fire on the burning platform. Visible direction from top leadership for support of the CI program is required.

Clarity of roles and responsibilities is the next critical success factor in a CI program. We'll cover the specific roles and responsibilities of a CI program in the next section. But for now, it's important to understand why this is an important success factor. Think about it. Culture change, by definition, means behavior change. No one really wants their behavior changed by someone else. Most rational adults want autonomy of their own behavior. Clarity of roles and responsibilities helps to ensure that feelings aren't hurt, and therefore resistance isn't built, when the difficult work associated with affecting behavior change arises. Clearly communicating, maybe even overcommunicating, who is responsible for what, and why a change agent has a certain job and what that job is, helps the behavior change go down easier.

The third success factor of importance is a rigorous methodology. The methodology should be chosen in consideration of the business strategy and the burning platform. Many organizations that are in need of safety performance improvement choose behavior-based safety improvement (Byrd, 2007). Many organizations that are in need of cost control, productivity, or capacity improvements choose a Lean Production methodology. Organizations that are looking to improve customer satisfaction, product quality, and process capability choose a Six Sigma methodology (Favaro, 2015). Remember, the methodology is important; it's the fulcrum of the culture change. There are many other moving parts. You can't just go hire a group of external Six Sigma consultants to do a bunch of improvement projects and expect a culture change.

It's also important to note that the *rigor* of the methodology is important. An organization that tries to use a watered-down, easy, short-cutting methodology will not see a culture change. The difficulty in mastery of the methodology, builds the organizational strength required to solidify the culture change. The bigger the fulcrum the further the weight can move. A tiny little bump of a fulcrum will only allow you to move the weight a tiny distance. A rigorous methodology, chosen to be in alignment with the performance improvement needed, is a nonnegotiable factor in a CI program.

The next success factor is linkage to business strategy. The CI program requires the allocation of resources. These resources come in the form of people, time, and thus money. However, it is unethical for managers in publicly traded business to invest resources without an anticipated return to the investor. So the organizational support for resources must be inextricably linked to business strategy. If the burning platform has been set by a poor safety record and injured employees, then the business benefit of improving safety performance becomes the return on investment in the allocation of resources to the CI program. If the need for culture change has been identified by the pain of dissatisfied customers and product recalls, then the cost of the CI program can be negated by the benefits to profitability, customer satisfaction, and revenue growth. The resources allocated to the culture change must have a return on investment and a long-term strategic benefit that enhances business performance, and therefore, must be linked to business strategy.

Finally, critical to the success of the CI program is the ability to measure the outcomes associated with the culture change. We need to know if we're moving the culture as we attempt to move the culture. Measuring the outcomes of the culture change helps determine the appropriate allocation of change agent resources. If we divert change agent resources, will the emerging culture change reverse back to the original culture? We need to be able to measure the robustness of our new behaviors. This is more than just measuring the new safety, quality, or profitability performance. This means measuring the behavioral aspects exhibited by the new culture. This might be in the form of measuring safe behaviors through observation, it might be in the form of improved process capability, or it might be in the form of measuring the number of audits completed by front line supervisors on adherence to standard work. What does the new organization look like? What are the observable behaviors of the new culture? That's what we need to be measuring.

2.6 Typical Continual Improvement Program

Let's discuss a typical continual improvement program in a typical large organization. If we understand what is currently typical, we can better see the subtle, but important, changes that make a CI program as CISR program. Let's take a typical major, global corporation and understand what, who, and how the program operates. Let's study some of the important factors that make a program successful. And let's understand how success is measured. Later, we'll talk about the details of turning CI success into CISR success.

First let's start with the definition of "program." We define a program as a system of procedures or activities that has a specific purpose. This definition seems particularly appropriate since it includes the word "system." Continual improvement should be conceived of as a system. It is not a standalone department, or group of people, or methodology, or series of projects. It is a system. A program is a system (see Figure 2.3). Systems have inputs, transforms, outputs, and feedback loops. Within the transform, there are procedures and activities. The activities need to be completed by people. Hence, there is an organizational structure to be built to carry out the activities, follow the procedures, and achieve a transformation within the system. This is our definition of program.

A CI program is a system in an organization (or a subsystem within a larger system) with the specific purpose to achieve performance improvement. The term continual refers to the repeating pattern of the activities within the program to achieve performance improvement. Once one level of performance is achieved, activities to achieve the next level of performance improvement are initiated. And this never ends. Customers and stakeholders will expect ever-increasing performance. There are a lot of moving parts in a CI program. Let's study the important ones. In the following sections, we will discuss strategy and targets; these are the means and ends to the continual improvement program. We will discuss roles and responsibilities; the people doing the work to achieve improvement. And we will discuss methodology; this is the recipe of action.

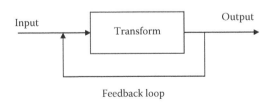

Input   Transform   Output

Feedback loop

Figure 2.3   A classical system.

The important elements of a CI program are

- Strategy and Targets
- Roles and Responsibilities
- Methodology

2.7 Strategy and Targets

A program of continual improvement must first determine what is to be improved. Most organizations have a set of performance measures that are of interest. A public, for-profit company may be interested in shareholder return or market valuation. A not-for-profit school may be seeking increased student enrollment or community reach. A government may be concerned with infrastructure strength or citizen security. A church might be focused on charity meals provided or senior citizen worship accessibility. Continual improvement programs are important to all types of organizations because all types of organizations are interested in performance improvement with respect to measures of interest. This type of forward looking interest of the organization is a strategy of improvement.

Many organizations also have improvement targets. In five years we want the organization to double in the number of charitable meals provided. Or, by next year we want a price-to-earnings ratio of 15.4. A city government might want to reduce the crime rate from 3300 incidents per year to 2500 per year. These improvement objectives are often referred to as "targets-to-improve." There is a specific measure, or metric, a time frame, and a performance target of interest to performance improvement.

A continual improvement program starts with a cogent understanding of the organization's strategy. Why are we here? What are we going after? What is our value proposition to our stakeholders? In service to that strategy there are measures of performance and achievement. For each measure, or metric, there can be a target for improvement, the next step in performance. The continual improvement program is built around this starting point.

2.8 Roles and Responsibilities

There are a few important roles, with each role having a set of responsibilities, which are of critical importance to a successful CI program.

These roles can be broken down into two general categories: generalists and specialists (see Table 2.3). And there are generalists and specialists at different levels in the organizations. Ultimately, the goal is that everyone in the organization is involved in the CI program in some way.

The generalists fall into the categories of champions, sponsors, and team members. Champions are the program supporters. These are typically top level leaders in the organization, and in the best case *the* top leadership of the organization. The champion's responsibility is to ensure the success of the CI program. Actions in support of this may include allocating human resources, capital resources, or creating and monitoring enterprise-wide measures of program success and support. Sponsors are the project supporters. Most CI efforts are done as projects, meaning there is a discrete start and a discrete end to an effort. Sponsors are those in a leadership position for the department or function interested in a specific project, a bounded effort of activities in support of specified performance improvement. The sponsor's responsibilities include ensuring that the project is funded and resourced, monitoring project progress, and identifying CI opportunities. Sponsor activities will include identifying projects, allocating

Table 2.3 Continual Improvement Program Roles and Responsibilities

ROLE	GENERALIST OR SPECIALIST?	RESPONSIBILITIES
CI program champion	Generalist	• Ensure the CI program has support in the form of funding and human resources. • Lead the connection between organizational strategy and CI results.
Project sponsors	Generalist	• Develop project charters by connecting performance targets to improve and project opportunities. • Ensure projects are funded with money and human resources. • Lead phase reviews to ensure project progress is meeting expectations.
Team members	Generalist	• Complete the process study and process change actions identified through the project. • Be an advocate for process changes. • Broadcast methodology tools and techniques.
Project leaders	Specialist	• Deploy problem solving and continual improvement methodologies to achieve improved performance. • Train and mentor others in the methodology. • Document and library lessons learned and problem solving results.

team members for a project, which includes freeing up time for effective team member involvement, scheduling phase reviews of projects to ensure timely progress and effective results, and making decisions on what is in and out of scope for a project. Team members are those people throughout the organization that are the "do-er's" of the process improvement. Their responsibility is to be engaged and participative in the problem solving and complete all the tactical actions necessary to implement the process change. The team members become advocates of the new process and broadcasters of the CI methodology. These generalists, at the top, middle, and lower levels of the organization are critical to the success of the CI program.

In addition to actively engaged generalists, successful CI programs have specialists. And, like generalists, how many specialists are employed is directly related to the size of the organization. These specialists often have odd names. For Six Sigma, specialists are called Green Belts (part-time practitioners), Black Belts (full time practitioners), and Master Black Belts (the top level of methodology experts). For Lean Production, specialists are called Sensei. This is a Japanese term meaning "teacher." Regardless of the name, these specialists are problem solving and process improvement experts. They have received specialized training in statistics, operations, team dynamics, root cause analysis, and other tools. The responsibility of the methodology expert and specialist is to deploy their knowledge for the benefit of the organization's performance, ensure the rigor of the methodology is effectively deployed, train and mentor others in the methodology, and ensure effective sharing and documenting of CI methods used.

2.9 Methodology

Beyond connecting the CI program to the organizations strategy for performance improvement, and providing for the necessary roles and responsibilities of organizational structural elements of the CI program, the final critical factor in a CI program is the methodology deployed. The methodology, or possibly methodologies, should be determined in consideration of the type of performance improvement intended. For example, if the business strategy is focused on reducing employee injuries then a behavior-based safety program might be an appropriate methodology. If the organization is seeking performance improvement in product quality, customer satisfaction, or process yield, then Six Sigma might be the preferred methodology. If the organization needs to strategically improve productivity, material velocity, or employee engagement, then lean production might be the chosen methodology. If breakthrough innovation is needed to stay competitive, TRIZ might become an adopted methodology. However, in many large organizations many different strategic opportunities exist and therefore the CI Program will adopt several methodologies simultaneously.

What is important to understand is that no single methodology is a panacea. The organization should use the knowledge of its problem solving and continual improvement experts to deploy the best methodology for the opportunity at hand. Understanding the limitations, pros and cons of different methodologies is needed. Marry the methodology to the opportunity.

2.10 Putting the CI Program Together

The CI program is sized to the needs of the organization. For a single office enterprise of less than 50 people, a single expert may fulfill all of the roles stated above. Large enterprises may have thousands of members devoted to continual improvement. Other very large organizations may have thousands of Six Sigma Black Belts. So, there isn't a single model for the organizational structure of a CI program.

In consideration of the size of the organization, the deployment of a CI program may go as follows. Top leadership challenges the organization to improvement performance with strategic objectives and targets to improve. Top leadership deploys a CI program by funding the initiative and hiring expert resources for program leadership. The experts identify and deploy the best methodology for the organization and its objectives. Training is done with champions and sponsors so that they understand their new roles and responsibilities. Training is then completed with new project leaders. A few high-impact projects are launched, mentored by experts, and closely observed by sponsors and champions. As results are witnessed and interest in the program increases, additional project leaders are trained and the portfolio of on-going projects increases. As performance improvements in broader functions of the organization are sought, additional methodologies may be introduced at this point in time. The program is sustained with the addition and promotion of program leadership and expertise. At some point, the organization reaches a saturation point with methodology expertise. The continual improvement methodology becomes an ingrained element of the organizational culture. Solving problems or reaching for performance improvement *without* the use of the methodology seems foreign. At this point, the culture change has happened. Top leadership reevaluates the CI program; the program's culture change goal has been achieved. The program's charter is changed to one of ensuring that new members to the organization are indoctrinated into the methods and culture; this usually significantly reduces the size and visibility of the program. The organization can now sustainably enjoy the performance level of the new culture.

Continual improvement is done to achieve an ideal state, be the topic quality, productivity, or social responsibility. But, the ideal state is never really achieved because the customer or stakeholders expectations are ever-rising. The purpose of the CI program is to shift the culture of the organization to one which will forever seek the ideal state with methods and expertise. The CI organizational structure is only needed *during* the culture change. Again, it may take 5, 10, or 20 years, but ultimately the end goal is to not have a CI program, because everyone in the organization is always devoted to continual improvement. The end goal of the program is culture change.

2.10.1 *The DMAIC Methodology*

Our method of CISR is based on the Six Sigma methodology. The Six Sigma methodology uses a five step process known as DMAIC (pronounced "dee-may-ick"): define, measure, analyze, improve, and control. So let's take a few pages and make sure that we all generally understand DMAIC. A common understanding of DMAIC will help when we begin to understand SOFAIR (pronounced "so-fair"). SOFAIR is the six-step process of achieving CISR. The six steps are stakeholders and subjects, objectives, function and focus, analyze, innovate and improve, and report and repeat.

A method is a body of systematic techniques used by a particular scientific discipline. The DMAIC method is a series of steps in the Six Sigma problem solving and continual improvement process. Within each step is a set of tools and techniques. Carefully following the process steps, and using tools and techniques within the process, is what yields the impressive results for which the Six Sigma methodology is so well known.

Many of the techniques used with the DMAIC method are not new. However, the application of the tools within a rigorous process is what yields improvements. Many people with a technical education, such as engineering, medicine, or research are taught "the scientific method." The scientific method is a process of the following steps, in this order:

1. Formulate a question
2. Develop an hypothesis
3. Make a prediction
4. Test the hypothesis through experimentation
5. Analyze the results
6. Make conclusions

The DMAIC method is similar to the scientific method. The DMAIC method is about 30 years old; the scientific method is over 300 years old. So what's the difference? The scientific method is embedded in the DMAIC method. However, there are some important book ends added. The scientific method as outlined above is very similar to the "measure," "analyze," and "improve" phases of DMAIC. It's the "define" and "control" phases that make DMAIC different. Seeking input, or the voice, of the customer is a critically important part of the define phase, and previous to DMAIC a forgotten step in problem solving. In the scientific method, the problem is identified and formulated by the scientist completing the study. In the DMAIC method, the problem is identified and formulated from the customer's perspective. We don't study what we want to study; we study what is important to our customer. And with the scientific method, once new knowledge is found and documented, assurances of a new standard is not necessary to the process. With the DMAIC method, significant energy is expended to ensure that new levels of performance are sustained. Thus, the control phase is missing from the scientific method.

Another peculiar, and effective, difference with Six Sigma from other methodologies are the "belts." In Six Sigma, there are Green

Belts, Black Belts, and Master Black Belts. Some organizations expand this to White Belts, Yellow Belts, Gold Belts, etc. But we'll just stick to the basic and universal three types of belts. Green Belts typically receive two weeks of training in DMAIC and basic statistical problem-solving techniques. Green Belts then complete continual improvement projects in the organization that are closely related to their daily roles and responsibilities. These Green Belt projects typically take between three and four months to complete. The projects are scoped small enough to be able to be successfully completed in this short amount of time. They are small scale, locally oriented performance improvement projects. Black Belts typically receive four to five weeks of training in DMAIC and advanced statistical problem-solving techniques. Black Belts complete enterprise-wide and large-scale projects. A Black Belt projects typically takes between 6 and 12 months to complete. These are complex projects. And the Black Belt has no other responsibilities other than completing Six Sigma projects. Master Black Belts are trained in advanced problem-solving techniques and are adept at advanced statistics. Master Black Belts often manage Black Belts and may lead the Six Sigma initiative for the organization in addition to leading very large scale, enterprise-wide projects.

Freeing up highly competent human resources to do nothing except continual improvement, as in the case of Black Belts and Master Black Belts, is an important innovation created through the Six Sigma methodology. Complex and difficult problems don't typically arise overnight. And they aren't solved overnight either. They take large amounts of data gathering, analyzing and experiments with cause and effect analysis, and then, possibly, significant policy and behavior changes to affect the performance improvement.

Another unique feature to the Six Sigma methodology is that of conceiving of quants of improvement in order to achieve continual improvement. The DMAIC method is a process completed within the boundaries of a project. And by definition, for projects there is a discrete start and a discrete end. The belts become project managers. Continual improvement is achieved through many consecutive small steps via projects.

As we reveal the SOFAIR method in the next section of this book, it is important to understand the distinctive elements of Six Sigma and its DMAIC method. The rigor of DMAIC ensures that problems are solved permanently, by expert resources, in an on-going basis.

- The DMAIC method is followed to ensure the right problems are solved permanently.
- Full time problem-solving professionals, Black Belts, lead projects.
- Continual performance improvement is completed through consecutive projects.

We will be using these critical elements for the benefit of CISR. SOFAIR is to CISR as DMAIC is to Six Sigma. Following the intent of the methodology is important; following the rigor of the method is critical.

2.10.2 A DMAIC Example

2.10.2.1 An Example: Define Phase

Let's go through an example to better understand the DMAIC method. Later, we'll solve the same problem using the SOFAIR method. You will be able to compare and contrast the two methods. Let's say that we want to improve the quality of tomatoes available at our local big box retail store. If you've ever sliced one of those pulpy, tasteless, slightly pink/slightly green things that are called tomatoes, in the winter…a big thick slice for your hot hamburger… you know that someone needs to solve this problem. Here is an opportunity for performance improvement. We'll step through the DMAIC phases here to show you how DMAIC works, and then we'll take the same example through the SOFAIR method later in the book.

First, we need to define our problem and scope our project in the define phase. We have a business case; we think that our store could sell 50% more tomatoes if they tasted like tomatoes instead of cardboard. Our problem statement might go like this, "In winter months, the taste value of tomatoes decreases from 4.0 to 2.2 (Likert scale) amongst Super-Mart shoppers. Tomato revenues decrease by 25% in the same months." Our goal statement might be to improve tomato taste from 2.2 to at least 4.0 year round. Those have to be some juicy tomatoes. We've just begun to charter our project. We also need to define who is on our team and what our milestones timing is for our project plan.

After the project has been scoped, we then need to talk to our customers. We might take a statistically significant sample of random shoppers and have them score their taste on various tomatoes. We would want to understand at what taste value they are compelled to buy more tomatoes. We would also want to understand if there are any other critical-to-quality (CTQ) characteristics, such as size, color, or firmness that might also be important to the customers' purchasing decisions. Through the project, we will want to ensure that all of the customers' CTQ's are either maintained or improved.

Lastly in the define phase, we need to make sure that the whole team understands our current process of vending tomatoes. Together, we will complete a supplier–input–process–output–customer (SIPOC) chart. This will give us a high level understanding of the process from farm to customer. And it will help us understand all of the input and output factors that might be important to study during the project.

We now have a goal. We understand that there is a business case to solving the problem. We know what our customers want. And we, as a team, have a general understanding of the whole tomato process as well as some inputs and outputs that, later, we might find are important. The define phase is complete when we have gained organizational support for the project, we understand our customers' expectations, and we understand what process we will be studying.

2.10.2.2 An Example: Measure Phase

Since we are a very large global retailer and we will want our solutions to be standardized at all stores, we will need to study a statistically valid sample set of stores. In the measure phase, we are most interested in understanding the current state. One issue is that winter happens in different months in the Northern Hemisphere versus the Southern Hemisphere. Different parts of the world have different tastes in tomatoes. Beefsteak tomatoes might be preferred in the United States, but small grape tomatoes might be preferred in Brazil. One solution to this extreme complexity is to pilot a solution in one area and then replicate the study (not the solution) in different geographical regions. So, we decide to revisit our project charter in the measure phase (that's perfectly acceptable in the DMAIC method) and rescope the project to only include western Canada in our initial study. This region has long winters and a love of tomatoes. If we can find an innovative solution here, it might help us study the problem in other regions.

Our first order of business is to ensure that our measurement system is valid, accurate, repeatable, reproducible, and stable. Unfortunately, our primary metric of success, customer subjective taste values on the Likert scale, is none of these. Luckily for us, our food scientists have already concluded that total sugar and sugar-to-acid ratio are predictive of customer taste. We will use a surrogate metric; and we will use a single food science lab in Vancouver for our measurements. We go through another round of testing to ensure that our total sugar and sugar-to-acid ratio measurements are valid, accurate, repeatable, reproducible, and stable.

Next, we'll take a statistically valid sample of tomatoes, over a determined baseline period of time, and measure our total sugar and sugar-to-acid ratio. We might note any other categorical information of interest during these measurements. This might include store information, tomato variety, farm source, tomato age, and package and storage information. We might also include information on other CTQ characteristics like color, firmness, and size.

Lastly in the measure phase, we want to determine the gap between our customer expectations and our current performance. We may want to go back to our original sample of customers and ensure that we have correlation between customer taste value and total sugar and sugar-to-acid ratio. And then, we can calculate the current process capability (the gap) between our ideal total sugar and the average and variation of total sugar in our baseline sample; we will also measure the gap between our ideal sugar-to-acid ratio and the average and variation of this ratio from our baseline sample.

In the measure phase we have narrowed the scope of our project with the intent to replicate our problem-solving methods in other regions later. We have chosen a measurement that can be tested for validity, accuracy, repeatability, reproducibility, and stability. We have taken a baseline sample of our study variable. And we have measured the gap between what the customer wants and what we have. The measure phase is complete when we have gathered baseline performance information on the current state, conducted measurement system analysis for data to be studied, and determine process capability for meeting customer expectations.

2.10.2.3 An Example: Analyze Phase

In the analyze phase, we are interested in understanding the relationship between input factors throughout the whole farm-to-customer process and the resulting taste. In this complex problem, we will have many input factors. All the input factors need to be prioritized for their contribution to taste. We can then manipulate the important input factors to achieve better taste.

So, the first order of business in the analyze phase is to determine all of the possible input factors of importance. We are not being selective yet. We want to know every possible input factor. We pull the whole team together, along with other tomato taste experts for a large brainstorming event. Collectively we come up with the following possible input factors: genetic variety of tomato, soil chemistry, fertilizer use, pesticide use, day planted, day harvested, harvesting method, ripeness at harvest, harvest duration, time between harvest and sale, truck temperature, store temperature, package type, display type, and 35 other factors. And each of these factors has multiple variables. For example, soil chemistry can be further parsed into pH, organic percentage, volume, and sand/silt/clay ratios. This is a very complex system.

Once we have all of the possible input factors of interest, we can begin to prioritize the relative importance of these factors. We are hoping that we conclude with two or three very important factors, such that when these factors are manipulated we always get great tasting tomatoes. For this step, we will use a prioritization matrix. This is an opinion-based method. With all of our experts together again, we ask them to weigh the relative importance of the output factors, or CTQ's. And then we ask them to rate their opinion of the relative importance of every input factor on the output factors (see Table 2.4). Through this technique, we find about four factors of interest.

Now that we have narrowed our list of possible factors, albeit through opinion not data, we begin to test the hypothesis, with data, that these factors are important. We will need to experiment. We will need to test different tomato varieties and the variety's impact on total sugar and sugar-to-acid ratio. We need to test different distances between store and field. We need to test ripeness at harvest and different time between harvest and sale. We have a lot of data to gather.

The analyze phase is complete when we have a full understanding of the influence of input factors to our output factors of interest. We

Table 2.4   Analyze Phase Prioritization Matrix

INPUTS/OUTPUTS	TOTAL SUGAR	SUGAR-TO-ACID RATIO	SCORE	RESULT
Weight	9	3		
Variety of tomato	9	9	$(9 \times 9) + (3 \times 9) = 108$	Very important
Soil chemistry	9	3	90	Important
Fertilizer	1	1	12	Not important
Pesticide	1	1	12	Not important
Ripeness at harvest	9	9	108	Very important
Harvest duration	1	1	12	Not important
Time between harvest and sale	9	3	90	Important
Truck temperature	3	3	36	Not important
Store temperature	3	3	36	Not important
Package type	1	1	12	Not important
.....				

call this the $Y = f(x)$ relationship, with Y being the process output, x being the process input, and f representing the function that relates the input to the output. Once we understand $Y = f(x)$, we can conclude the analyze phase. The analyze phase is complete when we determine the functions and relationships between our process inputs and outputs.

2.10.2.4 An Example: Improve Phase

For this example, in the improve phase, we will begin to formulate solutions based upon the information we gained in the analyze phase. We need to find the best solution to our problem. And this may mean testing many different solutions and choosing from the best. We don't want to reach for the first or most obvious solution. We do want to have the best solution.

From our work in the analyze phase, through our experimentation, we are focused on improving ripeness at harvest and time between harvest and sale. We've determined that the tomato variety is influenced by local tastes and in our future global solutions we want to be able to provide many different varieties to our customers. So we don't want our solution to be contingent upon tomato variety. We also recognize that soil chemistry will be very dependent on region. So we leave that factor alone for now. We want to see if we can achieve our total sugar and sugar-to-acid ratios by only influencing ripeness at harvest and time between harvest and sale.

Next, we gather a new team together for some more brainstorming. In the analyze phase, we wanted deep process experts involved in brainstorming possible input factors of importance. Now, we don't want deep process experts; these team members are too embedded in the current state. Now, we want team members from many different professions and we want to use creativity and innovation tools to get us "outside the box" in our solution generation. We want to generate as many possible solutions as we can; and then we'll use our prioritization matrix again to converge on the best solution. In Table 2.5, we see that the team has invented three possible solutions to improve our two input factors of interest. Again, this is opinion-based analysis. So we'll need to verify the relationship between the solutions and the assumptions with data.

The team has come up with several solutions that allow the tomatoes to ripen on the vine longer and to simultaneously reduce the time between field and store. They have chosen cost to implement (low cost being better) and improvement to taste as their criteria for "good" solution. Note the weighting; taste is a more important criterion than cost. The prioritization matrix shows their selected solution. They will allow the tomatoes to ripen in the field and then airfreight the tomatoes from the field to the store. It was the team's opinion that selecting only fields near stores may hamper tomato availability (tomatoes don't grow very well in Western Canada in January). The team did not feel that their alternative idea of shipping the whole plant to the store, although cost effective, would not be very effective with improving taste.

Table 2.5   Improve Phase Prioritization Matrix

SOLUTIONS/CRITERIA	COST TO IMPLEMENT	EFFECTIVENESS TO TASTE	SCORE	RESULT
Weight	3	9		
Airfreight tomatoes from field to store after full field ripening	1	9	84	Good solution
Procure tomatoes from 20 km or less between farm and store after full field ripening	3	3	36	Less desirable solution
Ship tomato plants to store to allow ripening on the truck during transportation	9	3	54	Less desirable solution

The team must now design and develop the details of the new air-freight process. And they must validate that the new airfreight process delivers the improved taste desired, while meeting or exceeding any other CTQ's of interest. This is a complex solution to the complex problem and a whole new process of tomato vending must be designed. The improve phase is complete when we have validated the new process parameters and implemented all process improvements required to meet customer expectations.

2.10.2.5 An Example: Control Phase

The purpose of the control phase of the DMAIC method is to ensure that the new and improved process sticks. We don't want our improvements to unravel the minute attention is redirected. We also need to set up monitoring systems if they don't already exist, to ensure that our performance improvement has been achieved and sustained. Additionally, in the control phase, we want to document our new knowledge and initiate the activities necessary to begin to replicate our problem solving in other regions. We also want to recognize all the team members who helped us complete the project and formally close the project with our champion, sponsor, and team members.

For this example project, we will need to ensure that our new airfreight policy is managed and maintained for the long term. This might include putting in place long-term contracts with the air freighters, setting up management structure for the air freight activities, and setting up systems for the on-going training of all team members involved. Our goal is to ensure that our process improvements are in play for at least the next 5 years or until another Six Sigma project works to improve our improvements. This is the time to install the organizational structures to sustain the gains.

In the control phase, we need to set up a monitoring system to ensure that our improved customer satisfaction is measured on an on-going basis. We might install random testing at our stores. For example, we might have a few tomatoes every week sent to our food science lab for total sugar and sugar-to-acid testing. We might plot the results of this testing on a statistical process control chart and define the responsibility of the produce manager to escalate problem solving if the results are not what is expected. We might also institute an on-going process of in-store customer taste testing of tomatoes to ensure that our ultimate project goal of a taste value of 4.0 (Likert scale) or better is sustained. The point of this monitoring is to ensure that the gains were made and sustained.

Lastly, for this project, we will thank all of our team members by sending them a case of juicy ripe tomatoes. We will communicate the results to the global organization through the employee newsletter. And we will advise management on our recommendation for the next region to be improved and work with that project team as they initiate their efforts to improve tomato taste for their customers. Our control phase is complete when we have made the new process settings standard practice; we are monitoring process inputs and outputs to ensure new performance stability, documented and shared lessons learned from the project, and recognized and disbanded the project team at project closure.

2.10.3 DMAIC in Summary

The Six Sigma methodology has taken the scientific method and enhanced it to create the DMAIC method. Since its invention in the 1980s, the Six Sigma methodology has been used by many organizations to improve performance on many different performance criteria. It is a rigorous, data-driven, statistically oriented problem solving and process improvement methodology.

Careful attention to the define and control phases expands DMAIC beyond the scientific method. Seeking process improvement that is focused on what the customer wants is an important differentiator. We are not looking to improve our scientific understanding for the sake of knowledge. This is performance improvement in the opinion of the customer. Seeking voice-of the-customer information in the define phase is a unique and important difference with DMAIC. At the other book end of the project, the control phase ensures that this discrete effort of improvement in the form of a DMAIC project is only one step in an improvement continuum. The minute we confirm a new level of performance, the DMAIC method advises us to initiate the next opportunity for performance improvement. It takes a DMAIC approach to achieve continual improvement.

As one DMAIC project leads to another, continual improvement is achieved. And, as this effort spreads across the organization in different functions, as more and more people participate in projects, as more and more projects are successfully completed over time, the culture of the organization begins to shift. Team members begin to identify opportunities for projects. People begin to think in a DMAIC way. Even for very small problems, after much practice, defining, measuring, analyzing, improving, and controlling become a way of working. People begin to ask for data. People begin to expect decisions to be made through the use of data and statistics and this becomes an important aspect to the organizational culture.

DMAIC is a step-by-step problem solving and continual improvement process. As we shift to continual improvement for social responsibility, having a foundational understanding of define, measure, analyze, improve and control will be important. Appreciating the rigor of a method of problem solving will be needed before we teach you about CISR and SOFAIR. Like DMAIC, SOFAIR is a problem-solving method and the rigor of this method can lead to accelerated performance improvement as associated with social responsibility.

2.11 It's Time to Get to Work

We are finished setting a foundational understanding of social responsibility and continual improvement, now it's time to get to work. You now have the basic elements to start CISR and SOFAIR. In the next chapters we will explain SOFAIR in detail and show you some examples of how the SOFAIR method works on many different topics. We will encourage you to just start somewhere. Keep in mind that your work will be *continually* improving. Once your first project is completed, there will be the next and the next and the next. The beauty of now viewing social responsibility through the lens of seven subjects and the many issues is that you have many options for places to get started. The depth of incremental improvement and the breadth of all seven subjects give us many manageable ideas for projects.

As you embark on your journey as a CISR practitioner, and examine the myriad ways your organization needs to improve social responsibility performance, you might become overwhelmed. We have discussed that continual improvement is the pursuit of an *ideal state,* and we acknowledge that this goal can never be achieved. Not only are stakeholder expectations ever-rising, but our world is constantly changing, from raw material availability, to cultural shifts, and environmental conditions. As a CISR practitioner, it is important to have an action-oriented mindset, rather than succumbing to the very human response of becoming overwhelmed, discouraged, or defeated. The project-oriented approach (quants of improvement), like that used in Six Sigma, helps to initiate manageable action without becoming overwhelmed.

It is helpful to look briefly back in time, and realize that others have paved the path for you. As discussed earlier, the Quality movement dates back to the early 1900s, the birth of the assembly line and the later stages of the Industrial Revolution. But keep in mind that as recently as the early 1900s, cities in Europe and the United States were overwhelmed by piles of vermin infested, rotting garbage, poor sewage disposal, undrinkable water, noise pollution and billowing smoke from factories, crowded housing, insufficient medical care, and the resultant high mortality. It is valuable for the CISR practitioner to study historical challenges, and how they were addressed in order to have an informed point of view about today's problems. Quite simply, life in the city stunk.

In the history of mankind, the idea of doing the right thing goes back to the dawn of civilization and the reality that sustaining a business, like Kongo Gumi, for almost 1500 years, has been proven. But the idea of social responsibility, with the complexities as we know it today, formed in just the past 150 years. Every region of the globe has faced social responsibility challenges at different times, in conjunction with industrialization. If you are working in a country like China or India, which has had increasing scrutiny in the areas of uncontrolled urban development, labor issues, factory safety, organizational governance, and the environment, the problems might seem too big to solve. You are not entering into uncharted waters. The Industrial Revolution gave birth to activists, who pressured city governments to take responsibility for collecting garbage and cleaning streets, build sewer systems, supply clean drinking water, regulate factory noise and smoke, ensure that housing was safe, build hospitals, and recruit medical professionals. From London to New York, conditions in factories were unpleasant and often unsafe, with inhumane work hours and practices such as child labor. Addressing those challenges was disruptive at best, and violent at its worst. Countries experiencing rapid industrial growth face the same challenges.

But now you have a method to address these, and other challenges. Remember the lessons from the Quality movement. A program is a system. And continual improvement, even in small amounts, yields drastic improvement over time. Ultimately, everyone in your organization should be involved in some way. A SOFAIR Project is not for the "lone wolf."

It is important to realize that social responsibility is multidimensional, and the way organizations address their obligation to improve their social responsibility position is just as important as the problem that are identified. The age of simply writing a check to be out of the line of fire is over. There is a new "license to operate" that transcends geography, governmental oversight, or certifying organizations. The court of public opinion has become more pervasive in the social media age. Whether the topic is safety, quality, or social responsibility, poor operating practices can quickly become ubiquitously known. A single on-line video of animal cruelty on a factory farm can produce outrage within hours.

Ignoring the obligation to focus on social responsibility can be a recipe for going out of business. We encourage organizations to understand that social responsibility is not a "bolt on" to their "real business." The best way to approach social responsibility is to integrate CISR into every possible area of operations. Start somewhere. Don't get overwhelmed. This is not a quick fix and just a new way of problem solving and performance improvement. It is as fundamental as the shift from inspectors at the end of assembly line to the use of statistical process control to yield Six Sigma quality levels.

We have discussed what a continual improvement program is, and specifically discussed Six Sigma as one such program. We walked through a DMAIC example. DMAIC is one problem solving and performance improvement method within the methodology of Six Sigma. In the next chapter, we will discuss the SOFAIR method, a key method within the methodology of CISR. And we will follow our tasty tomato example through the SOFAIR process to see how this method differs from DMAIC.

3

THE SOFAIR METHOD

Let's, now, understand the SOFAIR method in detail. The SOFAIR method consists of six phases of problem solving and process improvement:

1. Stakeholders and subjects
2. Objective
3. Function and focus
4. Analyze
5. Innovate and improve
6. Report and repeat

It is important to remember that the method is a process. Every step (or phase) must be accomplished. You cannot miss a phase and expect rigorous results. Not every project will be exactly like another. Some tools will be used in one project and not in another. However, the primary purpose of every phase must be visited. Just like Six Sigma, the secret to success is in following the rigor of the methodology.

In the following sections, we will discuss the project phases and some of the tools used in each phase. We do not have space in a single book to cover every possible tool and technique that could be used in each phase. We will thoroughly discuss some of the most commonly used techniques. Please seek a trained practitioner for help with your first few projects. At the end of this chapter, we will revisit the tasty tomato problem to see how SOFAIR approaches the problem differently than DMAIC. We will demonstrate the tools used and walk you through a whole project using this real-world example.

3.1 Introduction to CISR and SOFAIR

We've thoroughly explored the benefits of a continual improvement program, how they work, and the important bits and pieces in a Six Sigma context. Now, we will transfer these benefits to the topic of social responsibility by using the CISR®* methodology and the SOFAIR method. CISR is a program of social responsibility performance improvement; it has defined roles and responsibilities and success metrics. SOFAIR is a problem solving and continual improvement methodology used in a project-oriented context. SOFAIR provides a rigorous method. As an easy-to-learn set of tools, it helps to get everyone in the organization engaged. It helps to build a culture of social responsibility performance improvement. Again, SOFAIR is to CISR, as DMAIC is to Six Sigma.

3.1.1 CISR as a Methodology: SOFAIR as a Method

A method is a body of systematic techniques used by a particular scientific discipline. The method of achieving Six Sigma quality levels consists of a problem solving and continual improvement process following these steps: define, measure, analyze, improve, and control. Similar to this method of having process steps, or phases, SOFAIR denotes stakeholder/subjects, objective, function/focus, analyze, innovate/improve, and report/repeat. Each phase has tools and techniques to use. The process is used within a project framework just like Six Sigma. The projects are led by a methodology expert. Not all tools and techniques will be used in every project, but every phase of SOFAIR will be completed in every project. SOFAIR ensures that all of the important aspects of social responsibility performance improvement are addressed. It ensures that performance improvement is legitimate and validated.

In the next sections, we will discuss each of the phases in more detail. For now, let's focus on the overall CISR methodology. Like Six Sigma, there is a significant benefit to following a rigorous methodology. The method provides a means by which short cuts and pet solutions are avoided. Like Six Sigma, the CISR methodology is that of conceiving of quants of improvement in order to achieve continual improvement. Improvement is completed within the boundaries of a SOFAIR *project*. And by definition, for projects there is a discrete start and a discrete end. Continual improvement is achieved through many consecutive small steps via projects. And the identification of

---

* CISR® (sounds like scissor) is a registered trademark and can be used with permission for non-commercial use. Contact SherpaBCorp.com for permission.

projects happens in synthesis with the business strategy. As business targets-to-improve are identified with respect to social responsibility performance improvement, SOFAIR projects are launched.

3.1.2 CISR as a Program

Because of the similarities with Six Sigma, organizations that have a robust Six Sigma program will launch the SOFAIR methodology within the Six Sigma program. It only takes a little bit of additional training for Green Belts, Black Belts, and Master Black Belts to become skilled in the SOFAIR method. The Six Sigma program has a project portfolio. The Six Sigma program has resources allocated, trainers trained, and metrics of success being managed. Often launching CISR does not mean launching a new program. It simply means adapting and expanding the existing Six Sigma program to include the topic of social responsibility through the addition of SOFAIR.

SOFAIR and DMAIC are both effective continual improvement methods. However, the similarities end there. It is cautioned to not use DMAIC to solve social responsibility problems. We will follow our tasty tomato example later in this chapter and we will demonstrate how much more effective SOFAIR can be, over DMAIC, for social responsibility opportunities. However, the well-trained practitioner should be able to move between methods as the topic shifts from quality or productivity to sustainability or social responsibility.

Implementing SOFAIR through an existing Six Sigma program allows for flexibility in scaling. Very large, global organizations operating in hazardous conditions or hazardous locations may have many social responsibility opportunities. Small organizations operating in a mature society may have fewer risks and fewer opportunities. If the CISR program were standalone, there would be fixed costs for the program which might prevent the smaller organization from reaching for social responsibility opportunities. And the problems and issues may be so complex in the large organization that it becomes difficult to launch any new program and see fast results. As an add-on to the Six Sigma program, both of these conditions are remedied. For the small organization, if a productivity target takes precedence, then that can be the focus of Black Belts. When a labor practice issue arises, a Black Belt can be diverted to address this social responsibility issue.

And for the large organization, with many deployed Black Belts, the addition of a new methodology and new targets can be incrementally increased toward the topic of social responsibility.

CISR as a program can simply be additive to the existing Six Sigma program. If the organization does not have a Six Sigma program but wants to start addressing social responsibility opportunities, then CISR can be launched as a program in an identical fashion to Six Sigma, including training "belts" in both DMAIC and SOFAIR. Then the organization benefits from well-trained problem solvers who are able to tackle a myriad of topics. In either case, there is a well-worn recipe for implementing the program; the program can be scaled and focused on the important business priorities without adding structure or significant human resources.

3.1.3 CISR as a Culture

Continual improvement is done to achieve an ideal state, be the topic quality, productivity, or social responsibility. But, the ideal state is never really achieved because the customer or stakeholder expectations are always increasing. The purpose of the continual improvement program, be it Six Sigma or CISR, is to shift the culture of the organization to one which will forever seek the ideal state with methods and expertise. The CI organizational structure is only needed *during* the culture change. Again, it may take 5, 10, or 20 years, but ultimately the end goal is to not have a CISR program, because everyone in the organization is always devoted to social responsibility performance improvement. The end goal is the culture change.

This has been one of the key benefits of Six Sigma programs. Because of the proliferation of involvement by being a Green Belt, or project team member, a broad swath of the organization becomes involved and engaged in improvement. Over time, many people learn the method. They begin to "think" in a Six Sigma way. The same can happen with CISR and SOFAIR. With a deployment model that is identical to Six Sigma, or additive to an existing Six Sigma program, over time, many people in the organization will be exposed to the task of improving social responsibility. Many people will learn SOFAIR by leading teams or being team members. They will begin to "think" in a CISR way. This instigates a culture change.

It is important to note that outstanding results are only achieved through a rigorous application of the process; garbage in, garbage out. All of the aspects of the program must be implemented; and it is only through successful results, measured objectively, that the program will be sustained. And only after many years of sustained success will the method and the program lead to culture change. It cannot be understated. You cannot cut corners with the methodology. You must effectively apply the roles and responsibilities. And the results of the program must be measured. Then, through the continual application of the methodology by experts, performance is improved, business results are achieved, the business sustains the practice, and the practice becomes culture. Then, as stated above, the program can be disbanded as the end goal has been achieved. The end goal is not a CISR program; the end goal is a culture change.

3.2 Operational Definitions of CISR and SOFAIR Terms

Before we dive in, to understand exactly how CISR and SOFAIR work, let's take a few pages and make sure that we are on the same basis with some definitions. This is a sneak peek at the methodology. In the next chapter, we will detail the roles and responsibilities of CISR and the tools and techniques within the SOFAIR method. For now, let's define the elements of CISR: continual, improvement, social, and responsibility. And let's define the elements of SOFAIR: stakeholders/subjects, objective, function/focus, analyze, improve/innovate, and repeat/report.

3.2.1 Continual

We choose the term "continual," rather than the term "continuous," intentionally. W. Edwards Deming made a distinction between these two terms (Deming, 2000). He used the term "continuous" to refer to the perpetual and incremental improvements made over time. His use of the term "continual" consists of the meaning of continuous *and* breakthrough and revolutionary improvements. Continuous improvement takes our current performance and makes it better. Continual improvement considers breaking the current system in order to reinvent it to achieve a step-change in performance.

CISR is a continual improvement methodology. It can be used for incremental improvement, and it can be used for breakthrough improvement. In either case, the expectation is that the act of improving is always happening. The minute that one level of performance improvement is achieved, the next level of performance is sought. This expectation causes improvement actions to be systematized. It is never finished. Therefore, organizational resources, organizational structure, training and development, and roles and responsibilities in service to continual improvement must be generated and maintained.

3.2.2 Improvement

Improvement is the means to achieve the goal. It is not the goal. And it is more than just change. Improvement sets the expectation that the baseline or starting point is accurately measured, so that upon change proof of improvement can be made. Change and actions alone are not enough. The outcome must result in performance that is of benefit, specifically to stakeholders.

Throughout the CISR methodology, tests of improvement are embedded. Assurance against change that does not result in improvement is a critical aspect of SOFAIR. The whole methodology is conducted in a way that checks the actions and changes against the objectives of interest to the stakeholders. Only when changes result in benefits to stakeholders, in consideration of specific goals, can they be claimed as improvements. And the methodology deploys assurances that this improvement is stable over time. You must prove you improved and your improvements must stick.

3.2.3 Social

Social responsibility is an important term to understand. It is a term often used incorrectly and misunderstood. For some, the social in social responsibility hints at socialism. This couldn't be further from the truth. The term "social" in social responsibility refers to social structures and the power of society to both harm and help our environment. It is not a reference to the political genre of socialism.

Social refers to human interaction, collaboration, and behavior. As the top species in the food chain, with the ability to destroy ourselves and all other species, we hold an enormous amount of power. And none of us live in isolation to each other or the environmental system around us. A polluter in Africa can affect air quality in India. A military conflict in South America can affect food markets in North America. Space exploration from Russia can affect employment in Europe. How we behave connectedly is the meaning of the term "social" in social responsibility. Think psychology rather than politics. And it is only when we humans have our social house in order that we have the resources to apply the brain power to protecting other species and the rest of the environment, and thus ourselves, from our negative impacts.

3.2.4 Responsibility

Our definition of responsibility refers to an ownership or accountability that we assume, rather than a law or statute to which we are punitively monitored. Responsibility is voluntary accountability. Because of our role in society, we have not only inalienable rights but also inalienable responsibilities. Because of our role in society and the significance of our potential impacts on people, other species, and the environment around us, collectively we should carefully consider the potential of our actions to have a negative impact on those around us. Because of our role in society, we recognize that everyone else is like us and we all want life, liberty, and the pursuit of happiness. Therefore, when we behave with a sense of accountability and transparency to others, these underlying desires and our means to achieve them can be recognized by those around us. We are responsible for the impact of our decisions and actions on the larger system around us.

3.2.5 Stakeholders

Stakeholders are anyone or anything that can be helped or harmed by our actions. For a business, organizational stakeholders can be employees, customers, suppliers, company owners, the local community, and the local environments. Our actions are considered improvements in the eyes of our stakeholders. Stakeholders are the judges of benefit or harm. The SOFAIR method always starts with open, honest, two-way communication, or dialogue, with stakeholders.

The purpose of social responsibility is to serve stakeholders. It is important to note that the owners and/or shareholders of a business are stakeholders too. Social responsibility does not mean service to external stakeholders *instead* of service to internal or owner stakeholders. People, planet, and profit are considered simultaneously (Elkington, 1998). One group is not greater or smaller in priority than another. *All* stakeholders should be delighted.

3.2.6 Subjects

The "S" in SOFAIR has a dual meaning. It refers to stakeholders and to subjects. And the subjects are defined by ISO 26000. This is the most balanced and comprehensive set of topics of interest to social responsibility that we've found. There are seven subjects: organizational governance, human rights, labor practices, environment, fair operating practices, consumer issues, and community involvement and development.

The purpose of paying attention to all subjects is to provide a balance and a breadth of service to stakeholders. For example, the term "sustainable development" has been commonly narrowed to only consider construction and building in service to environmental considerations. This is a good focus for some social responsibility effort, but what about community development to provide for educated citizens to maintain and utilize the environmentally sound building for the next 1500 years? What about effective labor practices that ensure that the construction workers of the sustainable development project are treated fairly? All seven subjects should be considered and balanced in order for effort to be holistically considered social responsibility. Social responsibility is a broad set of subjects.

3.2.7 Objective

Objective refers to the organization's objective for embarking on social responsibility performance improvement. This objective is connected to business strategy. It is not intended that an organization seek social responsibility performance improvement that runs counter to its business interests. Both social responsibility and business performance can be achieved mutually. The SOFAIR method sets the expectation that very early on in the performance improvement process, the organization formalizes and documents the objective outcome that it expects to achieve by conducting the continual improvement work. What is the goal or purpose of the CISR activity? That's the objective.

3.2.8 Function

Determining where to start improvement activity can be overwhelming. There might be so many opportunities, or so many unknowns about current state performance that it is difficult to formulate discrete projects. The function phase of the SOFAIR process implements tools and techniques to prioritize. This helps to select the most important part of the organization in which to focus continual improvement projects.

Evaluating the functions within an organization or process begins the adoption of systems thinking. SOFAIR performance improvement is a method that seeks to improve aspects of the organization as a system. Inputs, transformations, outputs, and feedback loops are all important elements of a system. Recognize that every problem operates in a system and, therefore, every improvement opportunity should be approached as a change to the system. Understanding the functions within the organization allows us to apply systems thinking.

3.2.9 Focus

The "F" in SOFAIR has dual meanings. Once the prioritized function has been determined, then the process of focus is selected. The SOFAIR method deploys tools and techniques in this phase that help to identify the scope of the project. We understand which process steps we will study for improvement, and maybe more importantly, which process steps will not be of focus.

An important aspect of CISR is to recognize that social responsibility performance improvement happens in quants of effort as projects. The focus element of the SOFAIR method recognizes that for any given project only a small part of the system will be the focus for improvement. And then, project after project, incremental improvement builds to deliver organizational transformation and culture change. The focus of SOFAIR allows us to tackle manageable size opportunities for social responsibility performance improvement.

3.2.10 Analyze

The analyze phase in SOFAIR is the meat in the sandwich. There are various tools and techniques deployed in this phase. The goal of the analyze phase is to understand how process decisions, inputs, actions, and participants affect the social responsibility of outcomes. We analyze the relationship of inputs to outputs. The analyze phase is about thoroughly understanding all elements and interactions in the system that is the focus on improvement.

Understanding causal factors *before* changes are made is critical for robust improvement. Time and time again we've seen organizations jump to action before understanding the cause-and-effect relationships of systems. At best these unstudied actions deliver no change to performance; at worst they backfire and degrade performance. Part of the rigor of the SOFAIR method is the requirement to understand cause and effect before process change is enacted in order to ensure that performance improvement (not just process change) is implemented.

3.2.11 Innovate

Like the "S" and the "F," the "I" phase of SOFAIR has dual meaning. In some cases, incremental process improvement will not result in the performance desired. Therefore, innovation is required. An entirely new way of approaching the process is needed. The process must be intentionally broken and reset. The desired social responsibility simply can't be achieved with the current process. In this phase, innovation tools such as the TRIZ and other innovation techniques are deployed.

3.2.12 Improve

The improve phase of SOFAIR is where the action takes place. Once we understand the cause-and-effect relationships of our process, we can then begin to choose new inputs, eliminate wasteful process steps, or improve the performance of participants in our system. This is when we improve our process. There are many tools and techniques also deployed in this phase. The end result of the improve phase is an action plan, implemented, that yields more socially responsible performance.

Effective project management and communication is the purpose of the improve phase. Controlled changes, with thorough communication of all process operators involved in the change, are needed in order for changes to be adopted and behavior to be affected. The improve phases focuses on managing and controlling the process changes as they happen.

3.2.13 Report

Two fundamental principles of social responsibility are accountability and transparency. Because of this, there is an emphasis on publicly reporting results and performance. There are many avenues of reporting. This phase of SOFAIR focuses on reporting the benefits achieved through the CISR effort. Ideally, the organization is already reporting on sustainability performance; however, if not, this phase of SOFAIR becomes an opportunity to institute this practice.

Reporting can be internal, external, or both. Formal sustainability reporting can be separated from other business and financial reporting, or integrated. And there are many reporting frameworks that can be adopted. Reporting helps to communicate and instill accountability for the sustainment and continual improvement of social responsibility performance improvement.

3.2.14 Repeat

The "R" phase in SOFAIR has dual meaning; it stands for report and repeat. Now that our new, improved performance level has been reported, because this is a continual improvement methodology, we immediately launch into the next level of performance improvement. This next round may be focused on a different topic, but we never stop improving.

It is the adoption of a continual improvement methodology that relieves any single effort from perfecting performance. We find that many organizations never get off the starting block with social responsibility performance improvement because the problems are so complex; it is difficult to solve them with a single effort or solution. Untangling small pieces of problems with discrete but continuing effort can make solving complex and difficult problems possible.

3.3 A Walk through the SOFAIR Method

Let's walk through an example to put all the pieces together. Just to get our bearings straight, let's clear some assumptions. We are operating in an organization that is strategically devoted to CISR. Within the CISR program, there are organizational members filling different roles and responsibilities. The organization has developed a strategy on what is important to its social responsibility performance. The organization has allocated resources to become experts at leading social responsibility performance improvement. Those experts employ SOFAIR as a problem solving and process improvement method. SOFAIR is deployed through discrete projects. Each project is team-oriented, follows a rigorous methodology, and has a discrete start and a discrete end. Each project yields incremental or breakthrough performance improvement. And many projects, over time lead to excellent social responsibility performance and a socially responsible culture.

As an example, let's go back to our tomato problem. We only have pulpy, tasteless, slightly pink/slightly green tomatoes available for purchase at our big box store. We want to improve the taste of the tomatoes and therefore increase sales and revenue. In the DMAIC example in Chapter 2, we achieved some performance improvement on this problem through the Six Sigma methodology. The example solution was to air freight our tomatoes to the big box store in Western Canada. In the following sections, we will take the same problem and approach it from a social responsibility performance improvement viewpoint. We'll take our jumping off point as our DMAIC solution.

The DMAIC method has done a good job of finding a solution to improve the quality of the tomatoes. However, there might be some sustainability issues with air freighting tomatoes. We've just significantly increased the environmental impact of tomatoes sold and we've made our supply chain more complex. The following pages take us through this same example. However, instead of using DMAIC to solve our tasty tomato problem, we will now demonstrate the same level of problem solving using SOFAIR. This example will focus on a high-level overview, a demonstration, of SOFAIR. Later, in Chapter 4, we will go through each SOFAIR phase in detail, with descriptions of specific tools and techniques. Although, we have found a solution that significantly increased the taste, texture, and color of tomatoes sold through DMAIC, let's see if we can achieve the same quality results with a more socially responsible solution.

3.3.1 An Example: Stakeholders and Subjects Phase

The first phase of the SOFAIR method starts by developing dialogue with our stakeholders. We want to fully understand the problem from the stakeholder's perspective. But first we need to make sure that we understand who our stakeholders are. For this, we employ the SIPOS (suppliers–inputs–process–outputs–stakeholders) tool (see Figure 3.1). This gives us a list of stakeholders with whom we will engage. We want to ensure that we've heard from our farmers, our transporters, our customers, and our store employees to ensure that we have fully appreciated their opinions and interactions with this problem.

Second, in the stakeholders and subjects phase, we want to ensure that we've fully explored all seven social responsibility subjects from ISO 26000 associated with the problem. We don't want to only focus on consumer issues. That's what we did with the DMAIC example. And we don't want to only focus on the new environmental issues associated with our air freight solution. We also want to consider fair operating practices, organizational governance, labor practices, human rights, and community involvement and development.

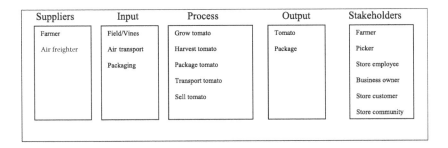

Suppliers	Input	Process	Output	Stakeholders
Farmer	Field/Vines	Grow tomato	Tomato	Farmer
Air freighter	Air transport	Harvest tomato	Package	Picker
	Packaging	Package tomato		Store employee
		Transport tomato		Business owner
		Sell tomato		Store customer
				Store community

Figure 3.1 The tomato SIPOS.

We want our new SOFAIR solutions to be as socially responsible as possible. We want to take a holistic approach to ensure that there are robust social and ecological solutions implemented. We want to make our customers happy with tasty tomatoes *and* we want to build great farming communities, have satisfied farm-to-table employees, promote fair competition for tomatoes in our market, ensure that our future tomato sourcing decisions are transparent and accountable, as well as ensuring that we are providing social systems that will ensure safe and productive ecological conditions for generations to come. We want tasty tomatoes for the next 1500 years.

3.3.2 An Example: Objective Phase

In the objective phase, we will begin to scope this specific SOFAIR project as connected to our overall business strategy. Creating a socially responsible supply chain for tasty tomatoes does not come at the expense of profit; it should enhance and build sustainable profit. Figure 3.2 shows an example narrative for this project. We want to make sure that we've stated our overall business strategy for social responsibility, make the business case for the project, effectively state the problem, and the goal for the project. We might include what is in scope and out of scope for this project. And we might document the team members and some information about the overall project management aspects of the project.

In this example, we see that our SOFAIR objective is linked to Value-Mart's overall business objective. The problem statement is still

---

Value-Mart Mission
It is Value-Mart's objective to deliver excellent products and services to our customers at a competitive low price. We will return value to our shareholders and be good corporate citizens in the communities in which we operate. We strive to achieve these objectives in a sustainable fashion.

Problem Statement
Value-Mart has recently become aware of substandard product quality of our tomato produce. Customers have complained about taste, texture, and color of our fresh tomatoes.

Business Case
Improving tomato quality, in a sustainable way, will lead to increased revenue, profit, and customer satisfaction, thus building employment security and community benefit.

Objective
Improve tomato taste, texture, and color while improving or maintaining sustainability.

Figure 3.2   The tasty tomato narrative.

oriented to consider the taste, texture, and color of fresh tomatoes. Note how the problem is not stated as a social responsibility problem, but as a customer satisfaction problem. The business case and project objective are stated as customer satisfaction problems. Customers are stakeholders. Improving product performance, tomato taste, is a social responsibility performance opportunity.

3.3.3 An Example: Function and Focus Phase

Now that we have a comprehensive understanding of our stakeholders' concerns across all seven social responsibility subjects, and we've connected the social responsibility performance improvement objective to our business strategy, we need to narrow our focus for our problem solving. It most cases, it is not reasonable to think that we will go from terrible performance to excellent performance in a single project. We need to understand what the biggest impact for our effort is in *this* project.

We will start with a value function diagram (see Figure 3.3). Here, we take the high-level process flow from the SIPOS and begin to do some high-level brainstorming, with our project team, on potential social responsibility failure modes at each process step. We will then

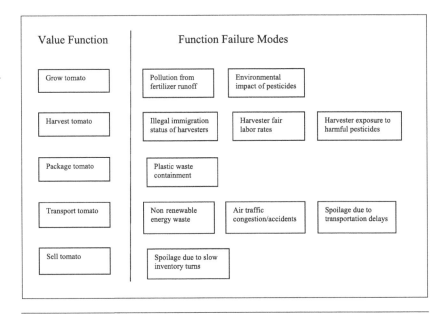

Figure 3.3   The tasty tomato value function diagram.

evaluate and synthesize our team, expert, and stakeholder input to determine the process step or steps that might have the most social responsibility risk. We will then narrow the scope of this project to focus only on those process steps.

The value function diagram delineates five general value functions: grow tomato, harvest tomato, package tomato, transport tomato, and sell tomato. This is a very high-level value stream map. Then, for each of the value functions, the team has brainstormed some potential social responsibility failures; these are the failures that are easy to identify. Again, a very high-level (not detailed) approach is acceptable. In this example, we see that harvesting and transporting the tomatoes seem to have many, easily identifiable potential failure modes.

3.3.4 An Example: Analyze Phase

The goal of our analyze phase is to understand the connections between our decisions, inputs, and resources and our social responsibility performance. The analyze phase is about understanding all of the complexities of the tomato supply chain system. Figure 3.4 shows work from a failure mode and effects analysis (FMEA) for our tomato problem. This is a risk assessment, leading to preventive action, for the highest priority risk. We analyze the severity and the probability of occurrence for socially irresponsible behavior across the seven subjects.

The FMEA helps us to understand that the air transportation aspects of the overall system are of concern with respect to the subject

Transportation FMEA

Subject	Failure Mode	Failure Effect	Severity Class	Occurrence Class	RPN
Organizational Governance	Unfair sourcing decisions for air transport supplier/corrupt sourcing decisions	Loss of investment return for business owners	7	3	21
Human Rights	No known impact				
Labor Practices	No known impact				
Fair Operating Practices	No known impact				
Environmental	Excessive use of nonrenewable resources for air transportation	Escalating fuel prices, air quality concerns, eventually unsustainable fuel sources	8	9	72
Consumer Issues	Unripe or spoiled tomatoes delivered due to transportation interruptions	Unsatisfied customers and lost revenue	7	6	42
Community Involvement and Development	Increases in air traffic congestion	Airport delays for local community members	3	5	15

Figure 3.4   The tasty tomato failure mode and effect analysis.

of the environment. It has the highest risk priority number. Also, the farm aspects of the system are at risk. If we can improve the social responsibility performance of both the transportation and farm issues simultaneously, we will have a social responsibility home run.

3.3.5 An Example: Innovate and Improve Phase

Now we thoroughly understand our complex system. We know what aspects of that system are at highest priority for improvement. Coupled with our stakeholder input, we have a rich body of information from which to begin to innovate and improve. Our goal in the innovate and improve phase is to invent many possible solutions and then choose the best amongst many.

This phase of our SOFAIR project proves to be tough. We need to leave tomatoes on the vine as long as possible. We need to shorten the duration of transportation between field and store. We want to improve the state of farm labor, and simultaneously improve our local Western Canadian communities. We want to make decisions that protect local ecologies and are transparent and accountable to our employee and business owner stakeholders. And we want to encourage fair market competition. There are some contradictions in the nature of this problem.

In this example, we use TRIZ to create solutions in consideration of these contradictions. See Figure 3.5 for our TRIZ contradictions and recommended solutions. We come up with the solution of building green houses on the roof of our big box store. We will grow our tomatoes, and other produce that is highly sensitive to transportation issues, on the roof of the store. Our cost is lower, our quality is better. We improve our local community by employing new "farmers" on our roof. The customer gets the very best tomatoes. We eliminate all ecological concerns from excess transportation. We recycle the heat and carbon dioxide from the store below. The SOFAIR process has given us an excellent socially responsible solution.

3.3.6 An Example: Report and Repeat Phase

After the roof-top greenhouse solution is fully implemented in our Western Canada region, we can begin to try this same solution in

---

Desired Outcome
We want to be able to pick the tomato after fully ripe and then rush it to the store for sale.
Therefore we want to increase the speed of the farm-to-table process.
At the same time we want to decrease the energy intensity of delivering the tomatoes, while
ensuring that the tomato is protected and stable during the delivery process.

Contradictions
Speed vs. decrease energy in moving
and/or
Speed vs. stability

Resolutions
Dynamics: Allow (or design) the characteristics of an object, external environment, or process to
change to be optimal or to find an optimal operating condition
and/or
Homogeneity: Make objects interacting with a given object of the same material (or material
with identical properties)

Solution
Change the dynamics of the transportation. Alter the tomatoes external environment. Make the
farm and the store homogenous. Put the farm on the roof of the store.

Figure 3.5   The tasty tomato TRIZ contradiction.

other parts of the world. We might have other types of produce in other parts of the world that are equally sensitive to ripeness, spoilage, and transportation concerns. We might build greenhouses for avocado trees in Mexico, or spinach in Poland, thus, tailoring the same solution for the local region.

We'll want to take credit for all of our sustainability performance improvement in our business reporting. Our owners and investors will want to know how much cost we are saving. Our employees will want to know how many additional direct employees are now engaged in growing produce and maintaining greenhouses. Our communities will want to know how much electricity and water we are saving. And our farm suppliers will want to know what new products they can now focus on for added value.

The report and repeat phase is about bringing this project to conclusion. We want to document and communicate what we've learned in this project through internal and external reporting. And then we want to take what we've learned and apply it to solve the same or similar problems in other areas of the organization in the spirit of continual improvement.

3.4 SOFAIR in Summary

This was a very high-level overview of how the SOFAIR method is executed. The output of each phase leads to the next phase. We do not jump to solutions; rather we spend a lot of time connecting our stakeholders' concerns to our business strategy. Within each phase, we use rigorous and proven technical tools to accomplish the work in that phase. We break the problem down and focus on the most important aspects. And we learn from the whole process.

We get a little better and solve one social responsibility problem in each project. But project after project we continue to get a little better and better and our social responsibility performance improves.

And we can see how much more socially responsible the SOFAIR solution is compared to the DMAIC solution. The DMAIC solution, air freighting the tomatoes from the farm to the store, solved a quality problem for customers. It solved a problem for one stakeholder on one subject. SOFAIR takes a more holistic approach and considers all stakeholders, internal and external, across all seven subjects. Yet the solution does not negatively impact business profits. The SOFAIR method is not about improving sustainability at the expense of the business. We want all stakeholders, including business owners, to be delighted for at least the next 1500 years. Putting greenhouses on the roof of our big box store is a solution that can grow and morph to solve a myriad of problems, sustainably.

You have also been provided an overview of a sampling of tools from which to choose. The process is to be followed diligently; phases cannot be missed. But the tools within each phase can be very simple and easy to use. Every tool does not have to be used in every project. Choose what makes sense and what fits the culture of your organization. You will be using a proven method to identify and prioritize risks and opportunities related to social responsibility.

4

How SOFAIR Is Deployed in an Organization

Now we are ready to teach you how to use SOFAIR and how to begin to lead your organization to social responsibility performance improvement. The following sections will explain each phase of a SOFAIR project in detail. We will highlight and demonstrate some powerful tools and techniques for each phase. This is not an exhaustive list of tools. There are many more tools we could have included in this book. However, we wanted to focus on the intent of each phase. If you are a Six Sigma or quality practitioner, once you know the purpose and goals of each phase, you can take the method and run with it. If you are a social responsibility practitioner, you can partner with the Six Sigma and quality practitioners in your organization to get their help to improve social responsibility performance.

This chapter will go through the purpose and goals of each stage of a SOFAIR project. Examples of analytic and problem solving tools will be demonstrated for each phase. Later in the book, we will show you four different projects completed with the tools used in each phase. With this information, you should have a good foundation to begin to apply the SOFAIR method in your organization.

It is important to remember the context in which CISR®* and SOFAIR are deployed. There should be full-time practitioners devoted to continual improvement. There should be a process that connects continual improvement efforts to business strategy. Then, targets-to-improve are developed such that measurable performance results are monitored and measured to ensure that performance improvement is connected to business improvement. In other words, SOFAIR (as projects) is conducted in consideration of an effective and integrated CISR program.

---

* CISR® (sounds like scissor) is a registered trademark and can be used with permission for non-commercial use. Contact SherpaBCorp.com for permission.

4.1 Stakeholders and Subjects Phase

We start our SOFAIR project by gathering information. The very first step of every SOFAIR project is engagement in dialogue with stakeholders. Performance improvement is defined by the stakeholder. Therefore, we start our performance improvement by understanding what constitutes *improvement* in the eyes of the stakeholder. It isn't improvement unless the stakeholder says it is improvement.

Engagement with the stakeholders, or gathering the "voice of the stakeholder," gives us the qualitative information to start the social responsibility performance improvement. It gives us a depth of understanding of the issues that need to be resolved. Analysis across the seven subjects of social responsibility gives us the breadth of understanding of issues that need to be resolved. Social responsibility isn't just environmental protection or consumer safety. It is a very broad set of subjects, and we want to make sure that we consider the full breadth of subjects for our performance improvement.

4.1.1 Stakeholders

We always start social responsibility performance improvement with stakeholder dialogue. Improvement is defined in the eyes of the stakeholder. Stakeholders are defined as an individual or group that has an interest in any decision or activity of an organization (ISO, 2010). For most organizations, there are hundreds, if not thousands or millions of stakeholders. We want to focus performance improvement on what is important to these stakeholders. But, with so many stakeholders, holding meaningful dialogue, understanding what they want, and aligning our strategies to their needs, is a complex undertaking. The tools we will use in the stakeholders and subjects phase will help solve this problem.

It is important to distinguish stakeholder dialogue from organizational monologue. We often see organizations push information to stakeholder groups through marketing, public relations, or social media channels. This might be important communication, but it is not stakeholder dialogue. Stakeholder dialogue involves listening. And ideally it involves listening while in the same place as the stakeholder, not through telephone calls, web surveys, or research reports. The best stakeholder dialogue happens in what is often referred to as "town hall" meetings. Representatives, with legitimacy, from stakeholder groups are physically present in the same place as organizational leaders. Discussions are facilitated such that crucial conversations happen without blame, argument, defensiveness, problem solving, or promises. The goal is to get information in the open. This approach takes careful training and expectation setting on the part of both the stakeholders and the organization. Problems are not solved during dialogue sessions. Information is gathered.

Table 4.1 shows an example town hall meeting agenda for one stakeholder group. Although most people represent multiple stakeholder groups (employees are also community members, suppliers can also be share owners); it is usually best to have the town hall meetings with only one stakeholder group at a time. This allows for more straightforward and focused dialogue. So, for a single SOFAIR project, information from as many town hall meetings as there are different stakeholder groups is needed. It is a daunting task. However, once this information is gathered, it can be used across many SOFAIR projects.

Table 4.1   Voice of Stakeholder–Town Hall Meeting Agenda

Agenda for factory neighbors town hall meeting	
Call to order	7:00 p.m.–7:15 p.m.
• Introductions	
• Roll call	
State of the business	7:15 p.m.–8:00 p.m.
• Historical performance	
• Market shifts and trends	
• Strategic plans	
• Near term objectives and challenges	
Floor open to facilitated conversation topic-by-topic	8:00 p.m.–9:00 p.m.
• Traffic	
• Parking	
• Noise	
• Employment	
• Trash removal	
• Water usage	
• Electrical usage	
• Operating hours	
• Other/new topics	
Meeting closed	9:00 p.m.–9:15 p.m.
• Review of record keeping	
• Next meeting time announced	

Face-to-face dialogue sessions are a great way to gather stakeholder information, but this isn't the only way. Data mining, surveys, focus groups, and benchmarking are additional ways to gather stakeholder information. However, these distant methods shouldn't be the only way to interact with your stakeholders. Also, it is important to remember that the organization is a stakeholder. Strategic information about the success of the business is important stakeholder information. Another important point is that stakeholder dialogue should not be a one-time event; there should a *system* of stakeholder dialogue, regardless of the method of gathering information.

4.1.2 Voice of the Stakeholder

After the stakeholder dialogue has taken place, and stakeholder information has been gathered, this information needs to be organized so that it is in a usable form for performance improvement. We use two tools to help us do this, the "SIPOS" and the "CTS tree"; both of these tools will be very familiar to the Six Sigma practitioner. Figure 4.1 shows an example of SIPOS and Figure 4.2 shows an example of critical-to-sustainability (CTS) tree. We will now explain these two important voices of the stakeholder tools.

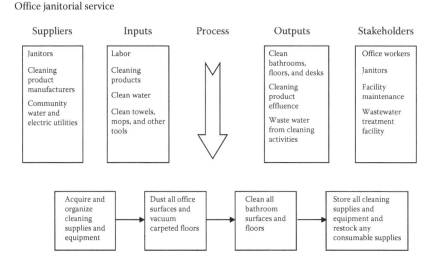

Office janitorial service

Figure 4.1   Suppliers—Inputs—Process—Outputs—Stakeholders.

Critical-to-sustainability
Process: office janitorial service
Stakeholder: janitors

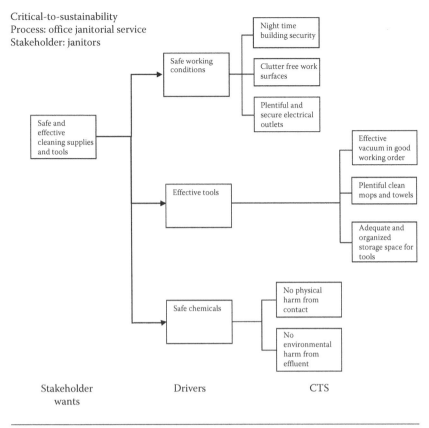

Figure 4.2   Critical-to-sustainability tree.

The SIPOS (pronounced sy-poss) diagram begins to apply systems thinking to our organization, see Figure 4.1 (Duckworth and Moore, 2010). We first outline the process that will be the study of the SOFAIR project. We draw a very high-level process flow diagram of this process, usually no more than three to five process steps. This is not a detailed process flow diagram. And most importantly, it will show us what we are studying and what part of the process we are not studying. The next step of completing a SIPOS diagram is to list the stakeholders of this process. For each stakeholder, there is an output of the organization that may have an impact on that stakeholder. Once the stakeholders are listed, think about the decisions and activities that are important to each stakeholder. These are listed in the output column. Then, we list all the inputs to the process and the suppliers of that input. The SIPOS is completed as a team exercise with all members of the problem solving team. This ensures that the whole team has a common understanding of what will be studied in the project and the nature of the inputs and outputs of the process.

In Figure 4.1, the SIPOS diagram is indicating information for the supply of janitorial services. The first step, in this example, is to determine the process flow for the delivery of these services. There are four high-level steps: acquiring the cleaning supplies, dusting surfaces and vacuuming carpets, cleaning bathroom surfaces, and then storing and restocking supplies and equipment. Note the boundaries of the process. It does not include the decision of whether to insource or outsource the service on the front end; it does not include the activity of maintenance or repair work. Once the process flow is determined (including where our study will start and stop in the process), then the stakeholders are listed: office workers, janitors, facility maintenance people, and our community wastewater treatment facility. We then list the outputs that impact these stakeholders. Finally, we list the inputs and suppliers of those inputs. In this case, the janitors are both suppliers and stakeholders. This is common that stakeholders can be suppliers to a process that is impacting them. The SIPOS gives us an indication of what will, and won't, be studied, who our stakeholders are to engage in dialogue, and some of the systems thinking needed to improve the process.

Figure 4.2 shows a critical-to-sustainability (CTS) tree. This is similar to the Six Sigma critical-to-quality (CTQ) tree. The purpose of this tool is to translate stakeholder wants into measurable outcomes that can become the focus of performance improvement. The SOFAIR project is not completed to achieve action without improvement. Therefore, we must be able to define and measure performance, and thus performance improvement, from the start. There is a bit of work involved with translating what the stakeholder said they needed in the dialogue sessions into performance outcomes. And once the outcomes have been established, they should be tested with the stakeholders.

In the example in Figure 4.2, we have completed the CTS tree that is a companion to our SIPOS from Figure 4.1. When creating the CTS tree we start on the far left with the stakeholders' general desires. In this case, considering the janitors as stakeholders, a critical sustainability driver is safe and effective cleaning supplies and tools. However, this is a vague and unmeasurable desire. We need to study this desire in more detail. We can parse this general statement to mean: safe working conditions, effective tools, and safe chemicals. And then for each of these needs, we can find measurable critical-to-sustainability factors. For example, safe chemicals are now defined as having no physical harm from contact and having no environmental harm from effluent. The human and environmental toxicity of the chemicals can be quantitatively and qualitatively measured. The mark of a good CTS characteristic is that it can be measured.

4.1.3 Subjects

Immediately after we have gathered and sorted stakeholder information, we begin gathering information on social responsibility subjects. The stakeholder dialogue tells us what is of interest to the organization's stakeholders. The review of subjects advises us on the breadth of subjects and issues of which we should be aware. This first phase of SOFAIR will culminate in an analysis that compares stakeholders to subjects. At the end of this phase, we will be able to begin our social responsibility performance improvement knowing what our stakeholders need across a broad range of subjects.

Social responsibility is defined in ISO 26000:2010 (ISO, 2010) as "the responsibility of an organization for the impacts of its decisions and activities on society and the environment, through transparent and ethical behavior." Along with this definition, there are seven core subjects. It is the breadth of these subjects that also help to define social responsibility. It isn't just environmental sustainability. It isn't just human rights. It covers every topic, with respect to the impacts of decisions and activities on society and the environment, for *all* seven subjects. These subjects are intended as a foundation of understanding for social responsibility. And within each subject, ISO 26000:2010 details issues, potential failures, or risks (see Table 4.2). Although each of the subjects was defined in Chapter 1, we will take each of these core subject areas and briefly describe the issues that should be examined, for each subject, within a SOFAIR project.

4.1.4 Organizational Governance

Transparent and ethical behavior is a part of the definition of social responsibility. Accountability, transparency, ethical behavior,

Table 4.2   ISO 26000 Subjects and Issues

ORGANIZATIONAL GOVERNANCE
- Transparency
- Ethical conduct
- Legal compliance
- Accountability to stakeholders

HUMAN RIGHTS
- Due diligence
- Risk
- Complicity
- Discrimination
- Resolving grievances
- Civil and political rights
- Economic, social, and cultural rights
- Rights at work

LABOR PRACTICES
- Employment relations
- Work conditions
- Social dialogue
- Health and safety
- Human development

ENVIRONMENT
- Prevention of pollution
- Sustainable resource use
- Climate change mitigation and adaptation
- Protection and restoration of the natural environment

FAIR OPERATING PRACTICES
- Anti-corruption
- Responsible political involvement
- Fair competition
- Using a responsible sphere of influence
- Respect for property rights

CONSUMER ISSUES
- Fair marketing, information, and contracts
- Protecting consumer health and safety
- Sustainable consumption
- Consumer support
- Consumer data protection and privacy
- Access to essential services
- Education and awareness

COMMUNITY INVOLVEMENT AND DEVELOPMENT
- Community involvement
- Social investment
- Employment creation
- Technology development
- Wealth and income
- Education and culture
- Health
- Responsible investment

respect for stakeholder interests, respect for the rule of law, respect for international norms of behavior, and respect for human rights are the principles of behavior, applicable across all core subjects. The core subject of corporate governance acts as an umbrella to the other six core subjects. Organizational governance should be integrated through all aspects of an organization's social responsibility program.

Organizational governance speaks about how an organization makes decisions, including the decisions associated with how the organization is structured. The primary issues to be resolved are transparency, ethical conduct, legal compliance, and accountability to stakeholders. Transparency means that the issues, factors, and outcomes of decisions and actions are open for stakeholder scrutiny. Ethics and accountability are concerned with decisions made with fairness and equity. Decisions made for short-term gain, in secret, and for the benefit of some in the organization at the expense of harm to others, is the long-term result of corporate governance that is not socially responsible. Are there opportunities in the SOFAIR project to improve organizational governance?

4.1.5 Human Rights

All humans have a right to fair treatment, the elimination of discrimination, torture, and human exploitation. ISO 26000:2010 defines eight different issues surrounding human rights: due diligence, risk situations, avoidance of complicity, discrimination, civil and political rights, resolving grievances, economic, social, and cultural rights, and rights at work. Concern about how stakeholders and their families are treated with respect to their rights of fairness as humans is the intent of this subject. Examples of failures of social responsibility toward human rights might be a failure to test for discrimination through hiring practices, a failure to ensure that labor sources are not creating demand for human trafficking, or supply chain sponsorship of child labor or compulsory labor. The risks associated with the subject of human rights are primarily those which exploit the humans in the organizational system. Are there opportunities in the SOFAIR project to improve human rights?

4.1.6 Labor Practices

Labor practices, as a core subject in ISO 26000, are primarily interested in employment policies and working conditions. This subject builds on the subject of human rights, but specifically addresses the responsibility of the organization as an employer. And these responsibilities as employer extend beyond just the direct relationship with the employee himself or herself, but also to the families of the employee, and the local community. There are five issues of concern with respect to labor practices: employment relations, work conditions, social dialogue, health and safety, and human development. These issues deal with the social contract between the employer and the employee. Failures of socially responsible labor practices would be indicated by the lack of recognition of employees as key stakeholders for the organization. Recognizing employees as a stakeholder is the intent of the labor practices subject of social responsibility. Are there opportunities in the SOFAIR project to improve labor practices?

4.1.7 Environment

We often see environmental issues as the sole focus of sustainability programs. This is a very limited view. It is hard to care for the environment in the middle of a war zone. Social stability is a precursor to environmental protection. We think this is why the environment is just one out of seven subjects. Although nothing is sustainable without a healthy environment, a healthy environment is not sustainable without responsible social structures. In general, ISO 26000:2010 approaches environmental sustainability through life cycle management and efficiency. A "cradle-to-cradle" approach is advised; what happens to a product after the consumer has extracted his or her value is as important as the sustainability of resourcing raw materials. There are four issues identified for environmental sustainability: prevention of pollution, sustainable resource use, climate change mitigation and adaptation, and protection and restoration of the natural environment. Failure to protect and nurture environmental status is a failure of this subject. Sustaining an abundant ecology for future generations is the intent of this subject. Are there opportunities in the SOFAIR project to improve the environment?

4.1.8 Fair Operating Practices

The subject of fair operating practices is about building systems of fair competition. Preventing corruption and encouraging fair competition helps to build sustainable social systems. There are five issues identified with fair operating practices: anticorruption, responsible political involvement, fair competition, using a responsible sphere of influence, and respect for property rights. An organization that bribes political officials to influence legislature would not be operating in a socially responsible manner; whereas an organization that is involved in the political process to ensure fair and equitable laws, taxes, or oversight is helping to build a sustainable society. The intent of the core subject of fair operating practices is concerning how organizations treat each other. Are there opportunities in the SOFAIR project to improve fair operating practices?

4.1.9 Consumer Issues

Consumer issues are primarily concerned with product safety, quality, and information. Organizations need the trust of the consumer to sustain transactions with those customers. There are seven issues with this subject: fair marketing, information, and contracts, protection of consumer safety and health, sustainable consumption, consumer support, consumer data protection and privacy, access to essential services, and education and awareness. Dangerous design, poor product and service quality, and a disregard of the privacy of consumer data are all failures of this subject. Protecting the health and welfare of the consumer is the intent of this subject. Are there opportunities in the SOFAIR project to improve consumer issues?

4.1.10 Community Involvement and Development

Local communities, interwoven, create societies. Community involvement and development help to create sustainable social structures. There are seven core issues of community involvement and development expressed in ISO 26000: community involvement, social investment, employment creation, technology development, wealth and income, education and culture, and health. It can be overwhelming for many organizations to consider improving whole societies; but, the ability to enact development and involvement within a local community is feasible. And community involvement is the means by which social responsibility is deployed. The intent of this subject is to create healthy communities, thus ensuring a sustainable society. Are there opportunities in the SOFAIR project to improve community involvement and development?

This was a very brief overview of the seven core subjects of social responsibility as defined by ISO 26000:2010. We highly recommend that your organization's CISR leadership be thoroughly educated on all elements of ISO 26000:2010. It is the international consensus on social responsibility. We use the core subjects of ISO 26000:2010 as a framework to understand the breadth of social responsibility. In this SOFAIR phase, we use the voice of the stakeholder to begin to comprehend the full range of stakeholder needs; and we use the seven core subjects to begin to comprehend the full range of subjects and issues that should be considered in our project.

4.1.11 Materiality

Materiality is a term often used (and abused) in connection with sustainability. Originally, it was a term from the accounting profession and is defined by the U.S. Supreme Court. A fact is material if there is a substantial likelihood that the fact would have been viewed by the reasonable investor as having significantly altered the total mix of information made available (SEC, 1999). The Global Reporting Initiative (GRI) (n.d.) has picked up this term and adopted if for use on subjects of sustainability. GRI encourages organizations to focus on sustainability subjects that are material to the organization and to do a materiality analysis to determine what subjects are material to the organization.

In Figure 4.3, we show an example of a materiality analysis. In this example, we have a simple four-quadrant priority matrix. The horizontal axis denotes those subjects that are important to the owner stakeholders (e.g., shareholders in a publicly traded company), and the vertical axis denotes the importance of topics for all other stakeholders. The organization in this example has chosen

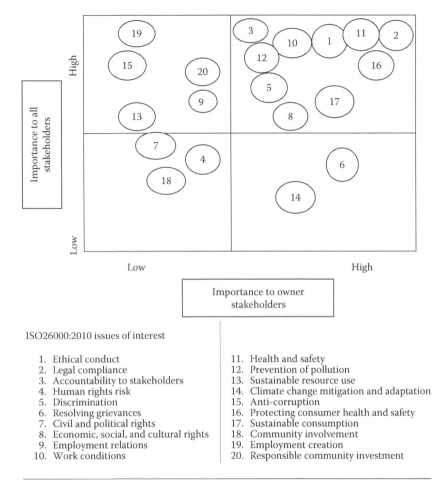

Figure 4.3   Materiality analysis.

20 issues from amongst the seven ISO 26000 subjects for analysis. Each issue is prioritized based on the two coordinates. Those issues that land in the top right quadrant should be pursued for continual improvement; these are issues that are important to all stakeholders. Issues that fall in the top left quadrant should be carefully evaluated. These are issues that are less important to the organization than other stakeholders. Why doesn't the organization recognize these issues as being material to CISR? Issues in the lower right quadrant (important to internal, but not external, stakeholders) could be investigated as potential communication opportunities of the organization to other stakeholders.

Every stakeholder has an interest in the success of the organization. These issues present an opportunity to ensure that all stakeholders are aware of these important issues for the organization. And issues that fall in the lower left quadrant can be deselected for CISR efforts. They are not a top priority for any stakeholder; we will improve them only after higher prioritized issues are resolved. This is an example of a very simple materiality analysis. It represents ways of determining what subjects, if not addressed, have a substantial likelihood to significantly alter the total mix of information shared toward performance improvement.

4.1.12 Stakeholders and Subjects

In the final tool that we will discuss for the stakeholders and subjects phase of a SOFAIR project, we put the two elements, stakeholders and subjects, together. Table 4.3 shows a Pugh Matrix, which compares each stakeholder with each of the ISO 26000 subjects. Not all stakeholders are interested in all subjects. And sometimes stakeholders have contradictory opinions on the same subject. For example, in Table 4.3, on the subject of community development and involvement, the local community and local wildlife are positively interested in this subject. The company shareholders are neutral; they care about community development if it doesn't take away from their profits. And suppliers and customers are concerned about negative impacts from the organization's community involvement and development; these stakeholders are concerned that activity by the organization may put cost and price pressure on them as the organization seeks to balance profits for the long-term interest of the community.

It is important, for many reasons, to understand which subjects are most important to which stakeholders. Communication about problems and successes can be targeted to interested stakeholders. Stakeholders with a high interest in a specific subject can be invited to participate on a SOFAIR project that focuses on that project. Metrics and measurements can be tailored to the stakeholder interest for specific subjects. Through SOFAIR we will be able to consider all stakeholder interest in all subjects. This type of analysis should be reevaluated periodically, but can be shared amongst many SOFAIR projects.

Table 4.3 Pugh Matrix of Stakeholders and Subjects

| | STAKEHOLDERS | | | | | |
SUBJECTS	EMPLOYEES	SUPPLIERS	CUSTOMERS	LOCAL COMMUNITY	LOCAL WILDLIFE	SHAREHOLDERS
Organizational governance	+1	+1	+1	0	0	+1
Human rights	+1	−1	0	0	0	0
Labor practices	+1	−1	0	+1	−1	0
Environment	0		+1	+1	+1	0
Fair operating practices	−1	+1	+1	+1	−1	+1
Consumer issues	+1	0	+1	−1	0	+1
Community involvement and development	0	−1	−1	+1	+1	0

4.2 Objective Phase

The purpose of the objective phase is to ensure that our social responsibility improvement efforts are connected to organizational strategy. We want to achieve the triple bottom line. It does us no good to take care of the planet and people while bankrupting the organization in the process. If there's no organization remaining, we can't be very helpful. We want to find ways to simultaneously have the planet, people, and the organization prosper sustainably for at least the next 1500 years.

We've finished the stakeholders and subjects phase. So, now we know what our stakeholders need across all seven subjects. We have a system of dialogue with our stakeholders. We have methods of considering a wide range of subjects. Now, in the objective phase, we synthesize the stakeholder and subjects information with business strategy. Our goal at the end of the objective phase is to have a concise statement of our mission, vision, goals, and objectives for our CISR efforts specific to the SOFAIR project being launched.

In the following section, we will discuss a short set of tools to help develop your SOFAIR project objective. We discuss business strategy as it relates to creating a social responsibility performance objective. We will use Hoshin Kanri (Babich, 2005), a form of strategy deployment. These tools are best used in the development of high-level organizational strategy. We will then discuss more project-oriented tools such as the project framework and project communications. Most large organizations have a strategic planning process. We encourage you to integrate your social responsibility strategy development with your organizational strategy development. In the end, we can't have social responsibility performance improvement without business performance improvement and vice versa. The objective phase is about developing narratives for our social responsibility performance improvement goals from organizational performance improvement goals.

4.2.1 Business Strategy

There are many books written on how to develop business strategy. Please refer to other resources if your organization is struggling to develop business strategy. In this section, we will focus on how to synthesize business strategy with stakeholder and subjects information to achieve a business strategy that considers social responsibility. This process, the process of synthesizing, preferably happens during the development of the business strategy. However, to begin with, assuming that social responsibility is not yet an integral part of your organization's business strategy, a process of synthesizing social responsibility strategy with an extant business strategy is needed.

The first layer of this synthesis considers the organization's business strategy with a 1500-year perspective. Who will be your suppliers, customers, and employees in 1500 years? Does the business strategy lead you to 1500 years of success? What assumptions of sustainable availability for suppliers, customers, and employees are in your business strategy? If any of these stakeholder groups die out (literally) what will happen to the business? Often applying this lens to business strategy brings sustainability and social responsibility gaps into focus.

The next layer of synthesis requires inquiry into the stakeholders and subjects information. Are there obvious gaps and dissatisfaction with one stakeholder group? How does the business strategy impact that stakeholder group? Are there social responsibility subjects that create business risks? Have these been considered in the business strategy? Ideally the business strategy is created with the stakeholder and subject information at hand. If the business strategy exists before the stakeholder dialogue takes place, then synthesis will be needed.

After these two layers of synthesis, it is important to recognize the business context as a sustainable system. Figure 4.4 demonstrates, graphically, how the organization's decisions, activities, and impacts exist in both the organizational context and the business environment.

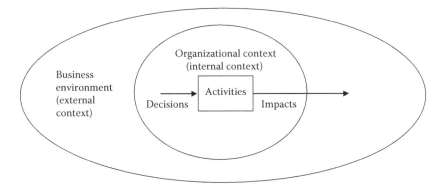

Figure 4.4   Objectives as a system.

Both organizational context and the business environment should be evaluated, especially as social responsibility shifts and trends may impact the context and environment. Are there pending changes to stakeholder expectations? Are there new competitors gaining sustainability advantage? Are there new regulations on the horizon? Is the organization growing in new regions? We find the strengths, weaknesses, opportunities, and threats (SWOT) tool to be useful in this layer of business strategy synthesis. Table 4.4 shows an example of a business strategy that is enhanced by evaluating organizational context and business environment subject to social responsibility, through a SWOT analysis. This example analysis is significantly abbreviated for demonstration. But, with just a little evaluation of stakeholder and subjects strengths, weaknesses, opportunities, and threats minor, but important, adjustments to the business strategy can be realized. Opening our view from product quality to delighting all stakeholders and from lowering cost to eliminating all waste, we have now created sustainability oriented aspirations.

Lastly, in order to synthesize business strategy with stakeholder and subjects information to yield a socially responsible business strategy, we suggest a process of appreciative inquiry. We want to see problems, potential problems, and risks as opportunities. We want our strategy and subsequent CISR objectives to be positive, to be proactive, and to be predictive of stakeholder needs. If your organization does not use appreciative inquiry, please refer to other sources to learn about

Table 4.4   SWOT Synthesis of Business Strategy

Existing business strategy: Regain lost market share by improving product quality and lowering cost.

SWOT analysis:

Strength	Weakness
• Global footprint	• Employee attrition is increasing
• Popular product	• Customer satisfaction is decreasing
• Technically talented organizational members	• Product redesign/updating is needed
Opportunity	**Threat**
• Growing product demand	• Primary competitor reports significant sustainability improvement through GRI G4 public reports
• Ability to simultaneously reduce cost and improve environmental impact	• Labor discontent at off shore factories
• Product redesign/updating is needed	• Raw material suppliers in coastal flooding impact zone

New business strategy: Regain lost market share by delighting stakeholders and eliminating waste.

this technique. For developing a business strategy that considers social responsibility, appreciative inquiry would have us answer positive-oriented questions. What does our company look like in 1500 years? What can we dream, when we dream big, for the future? How do we delight each stakeholder, group? What would our organization look like as perfectly sustainable? How can we translate our best performance to other areas or topics? Unfortunately, a lot of work around social responsibility has us look at threats and problems. Appreciative inquiry allows us to envision the best possible outcomes way beyond just solving problems or mitigating risk.

4.2.2 Hoshin Kanri

One method used to deploy business strategy throughout an organization is Hoshin Kanri. This Japanese tool is the basis for many continual improvement programs. Hoshin Kanri, also called policy deployment or strategy deployment, is a method of aligning strategy and improvement targets throughout the organization (Babich, 2005). Figure 4.5 is a graphical depiction of how Hoshin Kanri works. Starting from the top of the organization, the mission, vision, and business strategy are translated into key performance indicators and targets to improve.

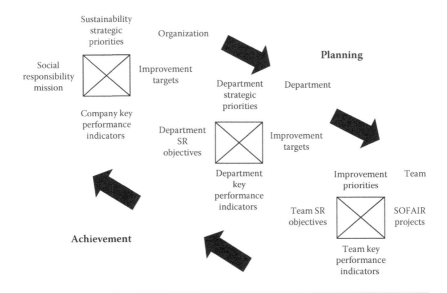

Figure 4.5   Hoshin Kanri for CISR.

These high-level improvement objectives are then translated to the next level, which might be group, division, or department. Each department then creates their targets to improve in service to the larger organization. Then, the department level is translated to team targets to improve, and so on, down to the lowest level of the organization. In large organizations, this is done with software to help monitor progress and assure alignment of performance improvement objectives from top to bottom. Ultimately, project portfolios are created; these are the actions that will be initiated to achieve the improvement targets. In our context, these would be SOFAIR projects, and each project can be linked all the way up to a top level objective.

Using Hoshin Kanri to derive social responsibility performance improvement targets from organizational strategy can be an effective tool in the objective phase. Because strategy deployment is parsed by organizational level and department or function, you can focus on just one or two departments to start your CISR objective development. You don't need to deploy Hoshin Kanri from top to bottom and wall-to-wall from the start. You may want to choose a social responsibility risk area, like procurement or operations and develop strategy deployment only for those functions. Or you may only want to use Hoshin Kanri for the top organizational level and one level down. This will still help you build objectives for performance improvement.

4.2.3 Projects

Once business strategy and targets to improve have been coalesced, now it is time to set some objectives at the project level. We won't necessarily have any specific focus at this point, that's the next phase of the SOFAIR project (function and focus); however, we can build some framework around a project. First, let's make sure we have some basic understanding of what a project is. A project is an initiative that has a discrete start and a discrete end; a program is an initiative that might go on indefinitely. A project has a single leader; there is one person responsible for the completion of the initiative. Projects are usually team-based efforts. The SOFAIR method is deployed as projects.

In the objective phase of SOFAIR, we can begin to formulate some of the information required for the project. We want to be as specific as possible while realizing that there are still many unknowns about

Table 4.5 SOFAIR Brief Narrative

Strategy	As an organization we believe that our strategy for success is to provide the highest quality product to a wide array of customers. The quality of our product is assured through our strategic focus on low cost and global appeal. This SOFAIR project is chartered to ensure that we consider the impact of our product assembly on all key stakeholders and that we become more a socially responsible organization.
Goal	The goal of the project is to identify the potential social responsibility risks association with the production operations for our products.
Business context	The context of this project includes the need to better understand the impacts that our product has on stakeholders and to determine if there are social responsibility risks associated with production processes.

our problem. Some of the best SOFAIR projects develop a short, one-page narrative that coalesces important aspects of the organization's social responsibility objective. Table 4.5 demonstrates a narrative. Developing a narrative as a project tool is a part for the failure mode effects and analysis (FMEA) technique (Duckworth and Moore, 2010). Often these narratives become living documents during strategy development efforts and can be recycled across many projects.

There are items in addition to those in the narrative that should be considered as project-oriented details in the objective phase. They include: who will be on the project, a time line for the project, and any tracking or numbering of the project, etc. Table 4.6 lists additional project charter elements (Project Management Institute, n.d.). This level of detail may or may not be necessary. It depends on the complexity of the project and the maturity of the CISR effort in the organization. Start with brief narratives if only a few SOFAIR projects are being attempted as a launch of CISR. If many project leaders are leading SOFAIR projects simultaneously, we recommend a more detailed project charter process; it will keep project leaders inside the boundaries of their projects and prevent overlapping efforts. In either case, a key part of the objective phase is to document the work plan for the project.

4.2.4 Communication

After the stakeholder and subjects information has been synthesized with business strategy to the point of detail that specific objectives for social responsibility performance improvement have been identified

Table 4.6   SOFAIR Project Charter Elements

Project leader's name
Name of person sponsoring the project
Name of person authorizing the project
Names of project team members
Names of persons who will approve the completed project
Budgeting information about the project
Start, end, and milestone dates
Project problem statement
Project business case
Project context or connection to business strategy
Stakeholder names or groups
Project goal
Expected tangible and nontangible benefits
Key performance indicators to monitor
Deliverables
Project scope, what is included and what is excluded
Project risks
The methods to be used
Key assumptions, constraints, or dependencies

and documented in a narrative and a project charter, this information should be communicated. To whom the project objectives are communicated depends on the project and the organization. It is our recommendation that all internal stakeholders be aware that a project is taking place and the project objective. However, there are risks with overcommunicating and communicating too early to external stakeholders. We don't want to get external stakeholders too excited or interested in positive changes to impacts before we are sure that we are going to work on certain impacts or before we are certain that we can positively affect the impact.

In some large organizations that will be managing many simultaneous SOFAIR project, internal websites are used to track and trend project status. In this situation, it is easy to pull key information from project tracking software for communication. During the objective phase, a broad communication on the goals and objectives of the project is desired. You may find organizational members, yet identified, who could be good team members on the project. You may find yet-to-be-identified internal stakeholders. There is little risk, and much to be gained, to communicating broadly internally as the SOFAIR project is launched. We leave the objective phase by communicating our synthesis of stakeholders and subjects information, the connection of project objectives to business strategy, and important details about the project being launched.

4.2.5 Function and Focus Phase

The purpose of the function and focus phase is to make sure that we are working on the most important opportunities. We will assume that your organization, like most, has limited resources and limited time to work on CISR. Therefore, our efforts need to be targeted on improvements that are important to the organization and important to stakeholders. Rather than have many people working on CISR all at once in different areas or on different topics (we call this the shotgun approach), we want to have an organized strategically targeted effort with resources working in concert to achieve objectives (we call this the surgical approach). With surgical precision, we want to apply resources to solve our problems and improve our processes.

Through the stakeholder and subjects phase, we understand what our stakeholders expect across a broad range of subjects. And through the objective phase, we understand what is strategically important to the organization. The function and focus phase continues with the synthesis that we started on the objective phase. We will further refine our synthesis so that we have a narrowly targeted goal for our SOFAIR project. This will require the application of systems thinking and prioritization tools. We want to consider the full life cycle of our products and services. We want to understand the functional aspects of our processes.

Stakeholders and subjects phase work considers the business environment outside the organization and bridges the external context of the organization to the internal context. Objective phase work studies the organizational context and strategies and bridges the external context, internal context, and organizational products and services. The function and focus phase bridges the internal context through products and services to processes. Performance improvement happens at the process level. See Figure 4.6 for a graphical depiction of the SOFAIR phases and how we arrive at targeted improvement. At the end of the function and focus phase, we will have a thorough

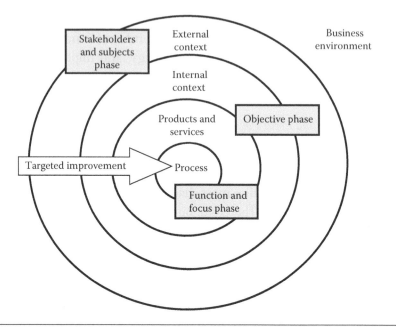

Figure 4.6   SOFAIR project preparation in concert.

understanding of the process that delivers the performance to be improved. Through this phase, we will develop precise goals for our SOFAIR project.

4.2.6 Function

The first half of the function and focus phase is interested in identifying performance improvement opportunities in consideration of understanding the organization as a system, and even systems nested within systems. In this phase, we will use life cycle analysis and process flow analysis as tools. The purpose of these tools is to develop an understanding of the opportunities and perils present in our processes, so that we can target the most important processes in the organization for improvement. In the function part of this phase, we look at the larger packets of work to determine opportunities at this level of granularity.

4.2.7 System Thinking

When beginning social responsibility performance improvement, many organizations struggle with where to begin. The issues seem overwhelming. Organizations are too complex. We usually don't have enough resources to address all the problems, much less the potential problems. One response to this is to scatter resources over several different issues (the shotgun). Another approach is to focus on the loudest or most persistent stakeholder issues, or even executive pet projects. Often these approaches do not have consistency of purpose over time; solutions tend to unravel after attention goes elsewhere. Or these approaches miss improving significant risks or valuable opportunities. They waste resources and poorly prioritize opportunities.

In classical systems theory, for every bit of work there is an input, the input is transformed through work to yield an output, and there is a feedback loop that signals the input on the status of the output, go back to Figure 2.3. If we begin to apply systems theory to our organizations, realizing that there are long chains of systems, parallel systems, and sometimes even contradictory systems, all yielding outputs of varying performance, we can begin to pick apart the complexity of the organization.

By applying systems theory to the functions of the organization, we can begin to identify areas of opportunity. We can consistently and cohesively develop targets for social responsibility performance improvement. We use a suite of process analysis tools to begin to expose performance improvement opportunities in different areas of the organization. These tools help us understand the interconnectedness of the functions and systems at play in the organization. From these process flow analyses, we will further refine targets in the focus part of this phase of SOFAIR.

4.2.8 Process Flow

The primary tools we will discuss are the life cycle analysis and process flow diagram. Other process-oriented analytic tools might be common in your organization. If your organization is familiar with value stream mapping, factory physics, function deployment flow charts, or other process analysis tools, use what is familiar to you. Your SOFAIR project may use some or all of these tools; you may even use other tools not discussed here, to better understand the chain of systems in your organization. The life cycle analysis can identify opportunities in all the steps in the lifetime of products and services. And the process flow diagram can be used for very coarse system analysis. Each will be discussed in further detail.

Life cycle analysis takes a look at products and services from the inception of design, through to resourcing raw material, making the product, consuming the product, and then disposing of the product (see Figure 4.7). If we do a life cycle analysis on an organization's products and services we will begin to see the connections between the impacts of phases of the life cycle. For example, let's consider an organization with a business of supplying software. Understanding the impacts of the design of the software would include all of the office space, commuting impacts, and hardware purchased for the design of the software. There are significant social responsibility impacts in the design of software. The primary raw material for the business of supplying software is the electricity to run servers and data centers. These resources have environmental impacts. Let's assume that our software company delivers its products virtually. The production and delivery impacts are embedded in internet access elements. Cell towers, cable lines and networks, satellites all have social responsibility impacts. The consumers purchase and use of the software might create identity security impacts. Data and privacy are an ISO 26000 consumer

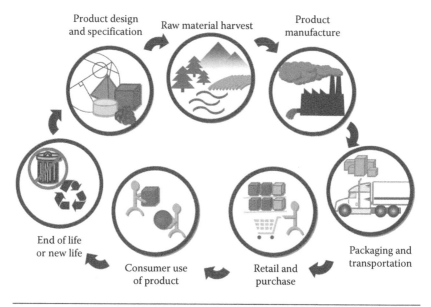

Figure 4.7   Lifecycle elements.

issue; data security is a social responsibility impact. Finally, the end of the usefulness of the software has social responsibility impacts. Replacement products for consumers, destruction of data, and financial impacts for upgrading to new software all have social responsibility impacts. Through this brief life cycle analysis example, we can see that something as seemingly benign as virtually produced software can have significant social responsibility impacts.

The life cycle analysis might point out a particularly impactful stage of the life cycle that we will want to focus on for process flow analysis. In Figure 4.8, we've shown an example of a process flow analysis for a packaging and delivery process. Here, we show every step in the process of packaging and transportation. We begin to see some of the complexity of the process. This is still a coarse, high-level look at the process. And the purpose is to examine the connections between the steps of the process.

Through the process flow analysis, we begin to apply systems thinking. We begin to see how precursors and antecedents lead to outcomes and impacts. If we focus on improving the inputs, rather than focus on controlling outputs, we will have more robust solutions. The further upstream in the process flow we can affect improvements the bigger the impact will be; the performance improvement can positively affect every process step downstream. We use a high-level process flow diagram, as functional analysis to see the process as a system.

Packaging and transportation

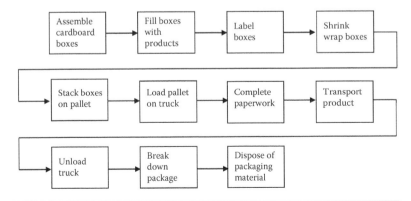

Figure 4.8   Process flow diagram.

4.2.9 Focus

Whereas the function part of the function and focus phase is intended to be a coarse view of the process flow by initiating systems thinking, the focus part of the phase dives into details. We will continue to work at the detailed process level through the analyze and improve phases. Now we will take the information we organized in the function part of the phase and begin to prioritize opportunities. We will use several tools to find the improvement opportunity that will give us the most improvement for the least amount of effort.

It is important to note that the best social responsibility performance improvement is preventive action. Our goal is to prevent irresponsible behavior before it happens. We hope that we don't have to correct irresponsible behavior after it happens. Therefore, when we start looking for improvement opportunities we are often looking for risks or threats. Not all risks are threats, and not all threats are risks. Risks are events that have a probability of loss and a probability of occurrence. And the probabilities may be so low that the risk of loss is acceptable; and sometimes risks carry reward. Risk, as risk, is not always to be avoided. However, we will use tools to analyze risks in order to find those with a high probability of loss and a high probability of occurrence, specific to socially irresponsible behavior. Threats have a known existence; the probability of occurrence is 100%. And, by definition a threat has a probability of loss, or negative outcome. However, because threats are known, preventive action can be formulated in advance.

The focus part of the function and focus phase is to find, in detail, threats and prioritized risks. We will then choose these risks to focus further preventive action in our SOFAIR project. In the hopefully rare case when we are working on corrective action to turn around behavior that has already gone wrong, the same prioritization tools can also help focus the project. In SOFAIR, we focus with the precision of a scalpel in the hands of a brilliant surgeon.

4.2.10 Qualitative Prioritization Tools

There are some prioritization tools that use qualitative aspects of analysis to determine priorities. We will demonstrate two tools in this chapter: value function diagram and quality function deployment.

Each of these tools uses qualitative data, usually the opinion of experts, to expose potential failures, risks, correlations, and priorities. Each of these tools has a lengthy history of use in the field of product and service quality assurance. Our instructions will be brief; we suggest you seek further instruction on the tool if you find that its use in your CISR efforts is warranted.

A very easy, simple tool to deploy for prioritizing potential failures is the value function diagram (see Figure 4.9). This is a simple brainstorming tool; it is usually completed by the team in a team setting. The example shown examines community access to mental health services. In this example, the team will first identify all the functions of access that provide value to the primary stakeholders. What are all the parts of access that are important to the stakeholders? Then, for each value function, the team will brainstorm potential failures. In this case, the team has examined, for each value function, ways in which community members may be prevented from receiving access to mental health services. This is not intended to be the list of all possible failures. These are just the failures the team was able to readily recognize. In this example, we see that follow-up, and monitoring of mental health treatment has the most readily identifiable failure modes. Therefore, this value function may be a good priority for the focus of the SOFAIR improvement. This very simple brainstorming

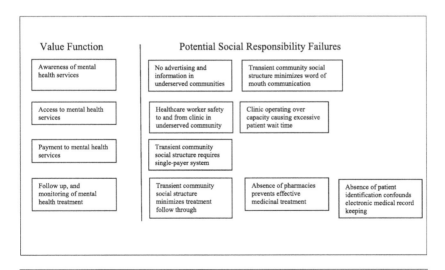

Figure 4.9   Value function diagram: mental health access.

tool has allowed us to focus on only one aspect of a large complex problem that will have the best impact for stakeholders.

The next prioritization tool we will discuss is the quality function deployment (QFD) matrix. This tool has historically been used to help product designers ensure that robust features are designed into products or manufacturing process that lead to fulfilling customer needs (Akao, 2004). Here, in Figure 4.10, we have continued the community access to mental health example. We have transferred this industrial design tool to the use of designing social responsibility aspects in the community. In this example, let's assume that the organization leading the SOFAIR effort is the local law enforcement organization. The police department sees the unfortunate situation of the incarceration of citizens in mental health crisis. They want to lead the resolution of this problem. The QFD matrix compares the needs of the stakeholders, in this case community members, and the needs of the organization, in this case the police department. We use the matrix to determine where shared goals and objectives meet; and, more importantly, the QFD matrix tells us where there are gaps in shared objectives. In this case, we see that patient containment and housing during treatment has a negative interaction with availability

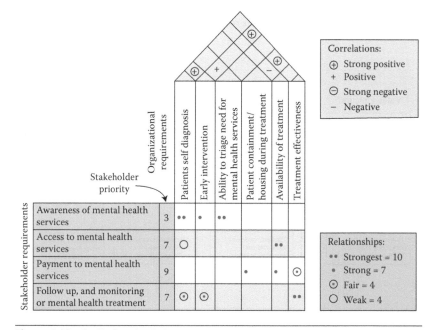

Figure 4.10 Quality function deployment.

of treatment, but a strong positive correlation to the effectiveness of treatment. And we see that patient containment and housing during treatment is not yet on the list of needs identified by the stakeholders. This is a gap that may be an opportunity for the SOFAIR project to solve. The QFD has identified a disconnect between the stakeholder and the organization; if this disconnect is resolved, then better performance by the police department and a better served community may result.

Although the value function diagram and the quality function deployment tools are qualitative tools, based on expert opinion rather than objective data, they can be very useful to help focus the projects. Understanding potential failures throughout the process flow through the value function diagram can show us a process step that might be more productive than others on which to focus the SOFAIR project. And the QFD can show alignment between stakeholder needs and organizational objectives, as well as interactions for all factors, thus highlighting gaps or opportunities with significant overlap. If these tools are not already in use in your organization, we suggest that you seek additional training, and use them for your CISR efforts.

4.2.11 *Quantitative Prioritization Tools*

We get heavily into data and analytics in the analyze phase. But, you might have some quantitative data already at hand; you can use this quantitative data to help you focus your SOFAIR project. We will discuss two quantitative prioritization tools: theory of constraints and the Pareto diagram. These are both simple tools that can be applied to sift and sort through different aspects of the opportunity in a way that allows you to focus on the most important and positively impactful continuous improvement effort. Again, if your organization uses additional quantitative prioritization tools, such as hypothesis testing prioritization matrices, Pugh analysis, analytic hierarchy process (AHP), or others, these too can be used in a SOFAIR project.

Figure 4.11 shows an example of using the theory of constraints (Goldratt, 2014) to focus improvements for a labor practices issue. In this example, we are concerned with worker safety from high temperatures in mines. The theory of constraints uses systems thinking to break the goal into success factors and the necessary conditions for

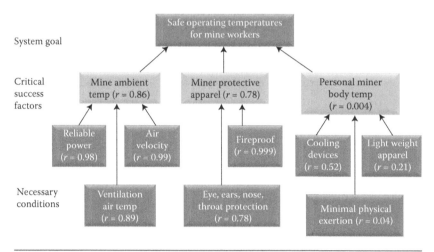

Figure 4.11 Theory of constraints.

those success factors. The theory of constraints tells us to focus on the constraints to our goals, and ensure that all surrounding processes are managed in service to the bottleneck, or constraint. Focusing improvement effort on a part of the system that isn't a constraint is a waste of time. Regardless of how good subsystems operate, the whole system will still be limited by the constraint. Only improvements in the constraint will yield improvements in the system. And constraints move. Once improvement to the constraint is made, a new part of the process may become the constraint. In the example in Figure 4.11, we see that there are three critical success factors to the safety of the miner: ambient mine temperature, miners' protective apparel (which has a negative, but necessary, influence on the miners' body temperature), and the miners' personal cooling ability. We have data on the reliability of each of the inputs to these success factors. Calculating cumulative reliability we see that we have good reliability in achieving effective ambient temperatures and personal protective equipment for the miner. However, our ability to successfully provide the miner with personal cooling ability is significantly lacking. The theory of constraints tells us that any work that we do on mine ambient temperatures, or reducing the heat load from personal protective equipment will not be as advantageous as improving the reliability of the miners personal cooling systems. We can now focus the project only on the personal cooling systems. Theory of constraints is a quantitative method; we need data to apply this tool effectively. But, if data exist, we can rapidly tighten the scope of the project and achieve social responsibility improvements faster and with fewer resources.

The next tool we will discuss is the Pareto chart. This is a simple way of arranging counts of categorical data. Pareto diagrams are bar charts with certain features. The bars are always arranged from most frequent to least frequent from left to right. The *y*-axis is always scaled to the total sum of all categories. And a cumulative percent line is drawn above the bars. See Figure 4.12 for an example. In this example, our organization has gathered data on the incidents of problems with legal work status amongst four different vendors. We see from the Pareto chart that Vend-Experts has significantly more work status issues than any of the other vendors. The Pareto chart tells us that over 60% of all work status issues can be attributed to Vend-Experts. In other words, if we focus our project on only Vend-Experts (doing nothing with any other vendor) we will solve 60% of our legal work status issues. The Pareto chart is an easy graphical technique; it requires data; but you can readily identify just a few categories that will make a big impact if they become the focus of your SOFAIR project.

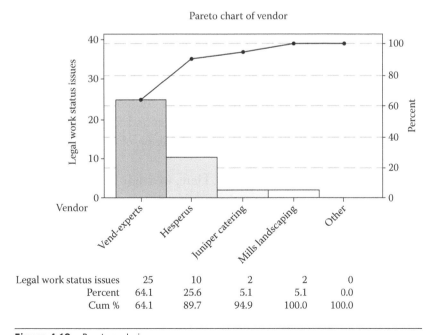

Legal work status issues	25	10	2	2	0
Percent	64.1	25.6	5.1	5.1	0.0
Cum %	64.1	89.7	94.9	100.0	100.0

Figure 4.12 Pareto analysis.

Our goal is to exit the function and focus phase with an understanding of how to focus the SOFAIR project. We now know, specifically, what to work on. And perhaps more importantly we now know what to not work on. Through the use of function tools, we begin to apply systems thinking to our opportunities. Through the use of focus tools, we use quantitative and qualitative tools to prioritize the opportunities. We can focus our team's efforts on the aspects of our CISR opportunities that will achieve significant performance improvement without a lot of resources.

4.3 Analyze Phase

The analyze phase is where we study the causal effects of our impacts. We want to understand the *relationships* between inputs and outputs. By this time in the project, we have received input from our stakeholders, and we have studied the issues of all seven subjects. We have an understanding of stakeholder expectations across a broad range of interests in the stakeholder and subjects phase of the SOFAIR project. We have synthesized the stakeholder voice with the organization's mission and goals to develop an improvement objective for the project. We now have a goal statement for the SOFAIR project through the completion of the objective phase. We took that objective and began to study the process of interest more thoroughly. We may have done some high-level research and evaluation on potential risks associated with different steps in the process. We might have gathered some data on existing issues and chosen to focus on one aspect of the problem. Through the function and focus phase of the SOFAIR project, we have a manageable scope of improvement for our project.

Now we begin the analyze phase. Here, we study in great detail the relationships between what we do and how it impacts our stakeholders. The analyze phase uses a large variety of tools; this phase holds the largest toolkit in the SOFAIR toolbox, in fact, too large to include all possible tools in a single book. In this chapter, we will discuss several tools of possible use. All of the tools we will discuss have their beginnings as quality improvement techniques. We have found that achieving the goals of satisfied customers and zero product defects is remarkably similar to achieving the goals of satisfied stakeholders and zero negative social and ecological impacts.

These tools fall into two categories: risk analysis and root cause analysis. For organizations improving social responsibility performance when not in a crisis state (good performance, moving to great performance) risk analysis tools are advised. This type of analysis focuses on preventive action, preventing problems before they arise and preventing irresponsible behavior before it is exhibited. For organizations in crisis (unacceptable performance, moving to acceptable performance) root cause analysis tools are advised. In this situation, we use tools that help us understand the system inputs that either allowed the irresponsible behavior to happen, or maybe even motivated the bad behavior to happen. In this type of analysis, understanding systems and antecedents is the focus. This is corrective action; we are correcting what has already gone wrong.

The goal of the analyze phase is to understand the process and system of study well enough to be able to design and implement improvements that are assured of preventing or correcting problems. There will continue to be unknowns after the analyze phase is complete. We will leave those for the next SOFAIR project in the spirit of continual improvement. However, for the small sliver of process that we are studying as defined in the function and focus phase, we want to understand the variable inputs, risks, and causal factors that could lead to irresponsible behavior.

4.3.1 Risk Analysis

Risk-based thinking is a critical tool in the SOFAIR methodology. Ideally, we are preventing irresponsible behavior long before it happens. Risk analysis requires us to understand negative impacts that *could* happen, in addition to understanding negative impacts that have happened. It is an exercise in pessimism. Our primary tool for risk analysis is the FMEA (Duckworth and Moore, 2010). This is a well-worn quality tool used during product and process design to ensure that product and manufacturing defects are prevented. This tool was originally created to ensure that military weapons systems achieved their kill missions (Military Standard, 1980). The irony is not lost on us that we are now using this tool that once was used for destructive purposes to improve the lives of people and planet!

The primary purpose of the FMEA is to identify the potential failures and their effects. The FMEA is a prioritization matrix documented as a spreadsheet, and the generation of potential failure modes is a brainstorming exercise. All reasonable, possible failure modes, for all seven social responsibility subjects from ISO 26000, will be considered in the FMEA. See Table 4.7 for a very simple example of a completed social responsibility: failure mode effects and analysis (SRFMEA). For the purposes of this book we will describe the main steps of completing an SRFMEA. Additional information on SRFMEA can be found in Duckworth and Moore (2010).

The first step is to brainstorm all possible failure modes as a cross-functional team exercise. This is tough to do; we are naturally drawn to claim what bad outcomes could happen; that's an effect. What we really want first is to understand what irresponsible behavior can happen; even with the best of intentions how can things go wrong (the failure mode), then we think of what the impact to society will be (the effect). We list all of the potential failure modes, or possible irresponsible behaviors. Then, we list all of the effects for those failure modes. There may be multiple effects for a single source of irresponsible behavior. Understanding the effects may bring up new ideas for potential failure modes. So we then circle back to the modes. Are there additional causes of failure to create the same effects? This is an iterative process: mode–effect–mode–effect–mode–effect.

The team may choose to work subject by subject for all seven subjects. For example, in Table 4.7 we've shown an FMEA for the function of the provision of new trucks for the service fleet of a company that maintains community electrical power lines. This is an abbreviated example; the actual FMEA might be 7–10 pages long. In this case, the team may choose to first tackle all modes and effects for "human rights," then all modes and effects for "environment," for example, continuing until all core subjects are covered. With this approach the team would first brainstorm all possible negative human rights impacts that could be caused by the purchasing new vehicles.

Next is the prioritization. There are two different ways to accomplish this task. The first is to rate effects, and you're always rating effects not failure modes, across the organization. One way to accomplish this is

Table 4.7 Failure Mode Effects and Analysis

FUNCTION	COMPANY TRUCK FLEET PURCHASE				
RESPONSIBILITY RISK	SOCIAL RESPONSIBILITY FAILURE MODES AND CAUSES	FAILURE EFFECTS	SEVERITY CLASS	OCCURRENCE PROBABILITY	RPN
Fair operating practices	Bribery from dealerships	Unfair competition between dealerships or brands	7	2	14
Labor practices	Inequality between employees in type of truck provided	Race, gender, age, etc. discriminatory practice perpetuated	3	4	7
Human rights	Vehicles purchased with components supplier nation with human rights issues	Unknown complicity with human rights violations through purchasing practices	8	7	56
Environment	Excessive fuel usage due to inefficient truck model chosen	Environmental pollution from fuel consumption	7	5	35
	Vehicle weight over-specified leading to fuel inefficiencies	Environmental pollution from fuel consumption	7	6	42
	Unreliable vehicle leads to early disposal	Excess material in waste stream	8	4	32
	End-of-life disposal is not easily recyclable	Hazardous materials leaching in landfills	9	8	72
Organizational governance	Decision to purchase new vehicles should not be needed if maintenance resources to older vehicles is provided	Environmental pollution from vehicle manufacture	7	6	42
		Financial penalty to company profitability for shareholders	5	6	30
Consumer issues	Unreliable vehicles cause customer service disruption	Safety critical services are disrupted	6	8	46
Community Development	No known impact				

to rank the effects within a single FMEA. The team gives the most severe effect the rating of 9, and the least effect the rating of 1. With this method, there should never be comparisons of one FMEA to another. The FMEA is a stand-alone analysis for the SOFAIR project. Now the failure mode and effects have been analyzed; however, the risk analysis is not yet complete. Some of these failure modes are more probable than others.

There are two factors to be considered when analyzing risk: the severity of the outcome and the probability of occurrence. We have just completed an analysis that rates the relative severity of outcome for many possible failures. Next, we will rate the probability of occurrence for the same risks. We strongly recommend that the rating of these two risk factors be completed in separate sessions with some time for reflection in between. Our experience has been that combining these steps leads to confusion and degraded quality of analysis. Turn back to Table 4.7 and study the occurrence column. Both severity and occurrence ratings are relative to each other and the opinion of the expert team assembled. It is common for teams to get too argumentative or detail-oriented on the application of ratings. It is important to remember that this is just a prioritization matrix to be used to determine which process improvements should be applied *first*. Don't lose friends over the completion of an FMEA.

The last step in the FMEA is to calculate the "risk priority number" (RPN), which creates the prioritization for action. The RPN is calculated through a mathematical manipulation of severity and occurrence. This can be achieved in two ways. We can either create a compound number from severity and occurrence. For example, a severity rating of 9, combined with an occurrence rating of 6, leads to an RPN of 96. Or we can multiply the two numbers together; in the same example the RPN would be 9 times 6, or 54. The purpose of the RPN is to highlight the highest priority opportunity for the application of preventive action. The former RPN calculation method always keeps our attention on the most severe failures. The latter RPN calculation balances the weight of severity and occurrence, but can allow for lowering priorities on potentially severe outcomes.

We have demonstrated the use of FMEA as a tool for prioritizing potential failures in social and ecological processes and systems. This is a risk analysis tool to be used by organizations using SOFAIR to move from good to great. This tool can be very effective when implementing new processes or products to prevent negative impacts. When continual improvement of social responsibility performance points to preventive action, we suggest using FMEA for risk analysis in the analyze phase of a SOFAIR project.

4.3.2 Causal Analysis

In the analyze phase, we use risk-based thinking to improve social responsibility performance by preventing irresponsible behavior before it happens. However, if irresponsible behavior has already happened, we will want to use causal analysis in the analyze phase. This type of analysis (also referred to a root cause analysis) is used to determine the process or system inputs that led to the negative outcome. There are many different tools to use for causal analysis. We will describe five common root cause analysis tools.

The first tool, "5-why's," may be the simplest, but often the most useful. This tool can be used by anyone, at any time, on any problem without any training. It is a great tool to spread widely and deeply throughout the organization. We start with a basic description of the problem as observed. We then uncover the immediate cause of that phenomenon by asking why it happened. Then we take the answer to the first "why" question, to ask why again, and again, at least five times. There is no magic in the number 5. It may take 3 "why's" or 13 "why's" to get to the root cause. If you start repeating your answer, or you arrive at a major systemic failure, as in the example, you stop the process. You know you've arrived at a root cause when you can back track, with evidence, by stepping back up from root cause to the observed issue with one step leading to the next.

Figure 4.13 shows an example. We start to solve a problem with security guards in the organization taking bribes from employees. Let's follow the "why's." When we ask why this happens, we find that this is common behavior for the location. The next "why" tells us that the subcontracting arrangement sets up the expectation for this behavior. The guards are paid below the minimum expected: their superiors expect them to be additionally compensated with bribes. Well, why did this happen? When the subcontract negotiations and communications were taking place, we were not explicit about the

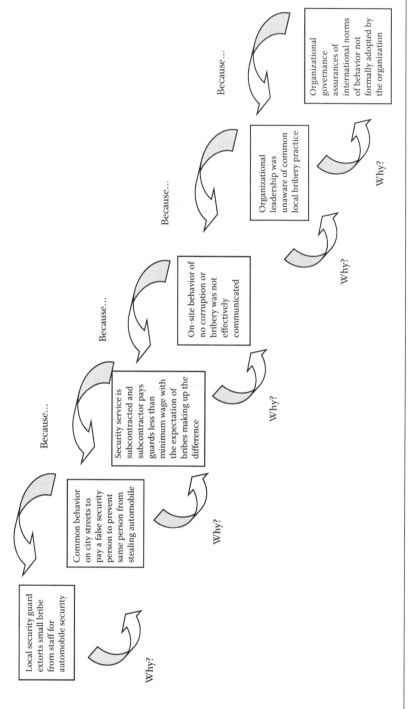

Figure 4.13 5-Why analysis.

expectations. And this happened because the direction from expat leadership, unaware of the local practice, was to seek the vendor with the lowest cost. Finally, after 5 "why's" we find a root cause. Our organizational governance assurances were not well deployed. We find a systemic social responsibility root cause. Rather than firing the vendor or punishing the guards, we should improve our organizational governance systems to ensure that *all* vendor selections and contracts explicitly disapprove of corruption of any kind.

The second tool is the fishbone diagram. This is a common brainstorming tool. It is useful when there are many possible root causes that need to be investigated and the next level immediate causal factor is not readily apparent. In this method, a team of experts and process operators are gathered together in a team environment. All possible root causes are brainstormed by the team. We don't yet know that these are root causes; it is the experts' *opinions* that these are possible root causes. Once the brainstorming is complete, a verification plan is then developed to test each possible cause in priority order. It is critically important not to assume that opinions are facts and to start changing the process based on the brainstorming from the fishbone diagram. A careful plan of root cause validation is needed after the diagram is created.

See Figure 4.14 for an example. In this example, we are studying the causal factors for the effect of diabetes disease escalation leading

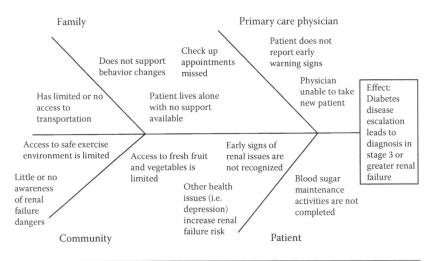

Figure 4.14   Fishbone diagram.

to a diagnosis in stage 3 or greater renal failure. We have four general categories of causation: family, community, primary care physician, and the patient. And for each of these categories there are several possible root causes that lead to our effect of interest. However, these are just possible root causes. The next step is that each of these possible causes must be validated to determine if the causal situation has created our issue. The fishbone diagram is a graphical tool used to organize brainstorming results.

The next tool is the fault tree. We recommend that the fault tree be diagramed by paying attention to both the conditional faults and the actionable faults (usually working in concert) that leads to the issue (Gano, 1999), see Figure 4.15. Each condition can then be further analyzed for the next level of conditions and actions. This tool can be used when multiple possible root causes are present. The best solutions are applied to conditions, not actions. If we improve the system to remove conditions that lead to negative impacts, we will have more robust solutions.

In this example (Figure 4.15), we start with the primary problem on the far left side of the diagram. Then we take this outcome and find pairs of conditions and actions that lead to this outcome. In this example, we are interested in the inadequate local candidate pool as we try to hire electricians for our growing business. We find the condition of a poor alignment with the local vocational school curriculum. Conditions are always existent, but don't necessarily cause a problem alone. It takes the combination of an action and a condition to cause a problem. In this case, with poor alignment with the curriculum the local high school students don't see the applicability and job prospects from the course of study. The root cause is an interaction between a poorly designed curriculum and student choices about the degree option. We can go another layer deeper on the poorly designed curriculum. There are conditions and actions at this level, too. Here, we find that the vocational school laboratory is outdated (that's a condition) and the partnered action is a poor partnership between the school and the company (that's an action). Our corrective action in this example would be for the business to sponsor an update of the lab to the technology needs of the business and help redesign an exciting course curriculum that leads directly to gainful employment. This

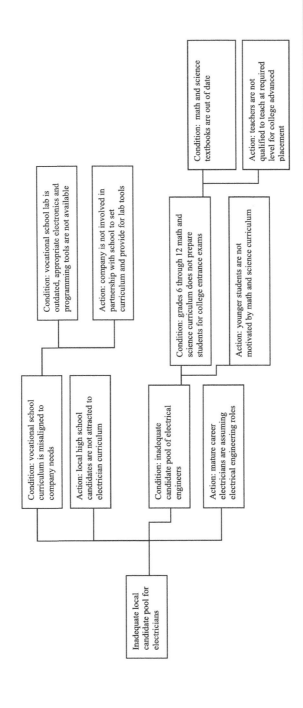

Figure 4.15 Fault tree diagram with condition/action analysis.

tool is a little more advanced than 5-why's and fishbone diagrams. We suggest additional training in this method.

The prior three causal analysis tools are effective when the phenomenon of study is qualitative. This happens primarily with human behavior or organizational systems having a relationship to the negative impact. The next two types of causal analysis are better used for issues caused by mechanical processes. They are hypothesis testing and regression analysis.

Figure 4.16 shows an example of a hypothesis test used in a SOFAIR project. In this example, organizational governance audit scores are analyzed by site. In this example, we use the analysis of variance hypothesis test; there are many different types of hypothesis tests to be used with different types of data to yield different types of statistics. From the analysis, concluded with the help of statistics software, we find that audit scores are statistically significantly different across the sites. We need to understand the audit process at sites A, B, and C and understand why the scores are different. Through the use of the hypothesis test, we can now focus our project on standardizing organization governance policies and procedures across sites. Hypothesis testing is an advanced statistical tool; training is needed in the use of this tool.

The last causal tool we will discuss is also an advanced tool: regression analysis. This statistical tool is used when a direct input factor can be correlated to an output measure. This tool requires that all inputs and outputs of study are continuous variables, or measurable numbers. In Figure 4.17, we show an example of using regression analysis to understand the risk in the current process of the potential of having effluent water released to a local stream. We have found a correlation between the pounds produced in our manufacturing process and the micrograms of effluent released. We can use the information from this analysis to either understand design parameters for new production processes, or to limit production pounds in the existing system. Additional training in regression analysis is recommended.

The purpose of the analyze phase is to thoroughly understand the relationships of process and system inputs to outputs and impacts before the process is changed. Charging into process changes before the cause-and-effect relationship is determined will waste time, resources, or possibly degrade, rather than improving performance. We will carry the information from this understanding of relationships into the innovate

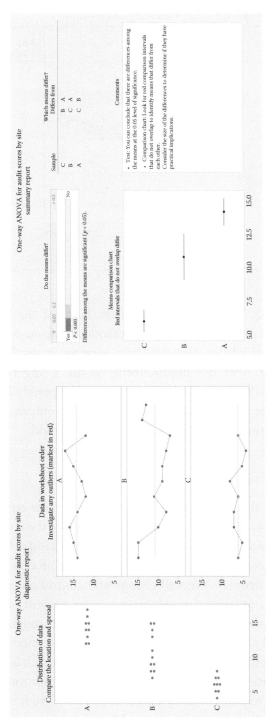

Diagnostic output from Minitab® software for analysis of variance (ANOVA) hypothesis test comparing organizational governance audit scores across three sites.

Figure 4.16   Hypothesis test.

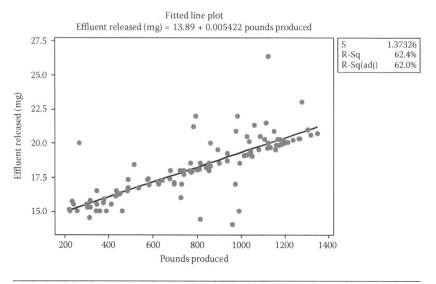

Figure 4.17   Regression analysis.

and improve phase. The analyze phase is the heart of the SOFAIR methodology; this phase separates this methodology from others that do not study causal relationships. We have seen too many "improvement" efforts that were not improvements at all. Using risk-based thinking and causal analysis to understand social responsibility issues as systems, and the relationships between inputs and outputs, elevates SOFAIR to a legitimate continuous improvement method.

4.4 Innovate and Improve Phase

After completing the analyze phase, we now know how our process inputs are related to our process outcomes. We can see the connection between our decisions and actions and the associated impacts. In the innovate and improve phase, we will put into place process changes to improve our outcomes. We will change and control our inputs to affect changes in outcomes through process improvement or we will create new processes with new inputs and outcomes through innovation. And as in all prior phases of SOFAIR, we will use methodological rigor to ensure that the process is only changed as much as needed, but changed intentionally and thoroughly to achieve improved results.

It is interesting to note that only after four phases of study and preparation we are now ready to take action. Many less rigorous methods start with changes. And, in many cases, these changes do not result in improvement. We have seen many cases where busy-ness and activity, rather than legitimate performance improvement, is a desired state in sustainability achievements. We challenge this assumption. We have seen many activities result in degraded, rather than improved sustainability. Only a rigorous method of study and understanding process relationships can we begin to understand the complexity of the social systems that we target for improvement. It is important to remember that change does not equate to improvement.

In this chapter, we will discuss both innovation and improvement. Innovation techniques are used to generate many possible solutions when the current process either does not exist (we can use SOFAIR preemptively) or the current process does not have the capability of achieving the performance needed. In the latter case, we deconstruct the current process and innovate to create a new process. Improvement methods are used in SOFAIR when we want to keep the current process but incrementally improve outcomes or mitigate risks. We will detail methods for validating that our improvements are improvements and methods for project management and implementation. Different projects will require different approaches; and some projects may require a combination of innovation and improvement. In either case, we encourage you to use rigorous tools of innovation and improvement. These scientific methods will help to ensure that your new and improved process performance is sustainable.

4.4.1 Innovate

We will use innovation tools to help us generate as many creative solutions to the relationships found in the analyze phase as possible. The goal is to have many possible solutions from which to select the best. We want to approach our problem solving creatively and innovatively. So, instead of picking the first solution that comes along, we want to motivate our team to create many possible solutions to the problem. This will ensure that the most obvious or easiest solutions are not selected; but, rather the best amongst many solutions is implemented. We will only discuss two innovation techniques in this chapter; however, there are many more innovation techniques.

The first innovation tool we will discuss is TRIZ. Again, TRIZ (pronounced like "trees") is a Russian acronym meaning "theory of inventive problem solving." This method was developed after World War II by Genrich Altschuller, a Russian inventor and naval officer (Altshuller, 1996). TRIZ is based on a theory that there are universal principles of inventiveness and creativity. It is a structured method that can be used to rapidly generate plausible concepts that solve problems in technical and nontechnical domains. TRIZ is based on the idea that many of the underlying "root" problems that engineers face contain elements that have already been solved, perhaps in a completely different industry, for a totally unrelated situation, that uses an entirely different technology. The root causes of dissimilar problems are remarkably similar. We can accelerate new problem solving by understanding how the same root cause was solved in a different situation.

The second innovation technique we will discuss is called 7-ways. This is a tool embedded in the production planning process, or 3P technique (Coletta, 2012). The 7-ways technique is a method of forcing the problem solving team to create at least seven different ways of solving the problem. The method also includes prioritization techniques to choose the best solution and two-dimensional and three-dimensional simulation techniques to build optimization into the design process.

If your organization is well practiced in different innovation tools, please use what is familiar to you. The purpose of the innovate and improve phase is to ensure that robust and creative solutions are found to social responsibility problems. We want simple, elegant, low-cost, robust, and sustainable solutions. Balancing these desires requires thoughtful and imaginative approaches. TRIZ accelerates innovation by investigating the connection between problems. The 3P and 7-way technique compels the team to evaluate at least seven different ways to solve the problem, thus seeking innovative solutions.

4.4.2 TRIZ

We start using the TRIZ method by recognizing two axioms: first, somebody, someplace has already solved your problem, or one very similar (creativity is finding that prior solution principle and modifying it to fit your circumstances), and second, don't accept compromises but rather, seek the ideal future result and remove the source of the problem. Also at the heart of the TRIZ method is the concept of contradictions. Altshuller recognized repeating patterns of contradictions and solutions to those contradictions. He built a matrix to pair contradictions and build recommendations for solution principles. We use this matrix of contradictions to understand a limited solution set of design principles (Solid Creativity, n.d.). There are many online resources that have documented the TRIZ contradictions matrix. The principles are then applied to find an innovative solution to our specific problem with a focus on using the resources that already exist in the system. Frequently, just understanding the contradiction can be a good step toward eliminating the problem.

Table 4.8 shows an example of an innovation process using TRIZ. The first step is to conceptualize the ideal future result. For example, we don't want a better chair, we want comfortable seating. The ideal seating system occupies no space, has no weight, requires no labor, requires no maintenance, and delivers benefit without harm. The ideal future result accepts no compromises. The next step is to begin to understand what is preventing you from achieving this ideal state. Go back to the analyze phase. What does the current system look like?

Table 4.8   TRIZ Problem Solving Process

PROCESS STEP	EXAMPLE
1. Formulate the Ideal Final Result (IFR).	Input mechanisms (buttons) on cell phones are the ideal size for all users.
2. Identify the problems inherent in the system that prevent you from achieving the IFR.	Some users want large buttons, other users want a small phone. Proliferating too many different products for different customers will increase costs.
3. State the problems as sets of contradictions.	Big buttons are ideal for some cell phone users. Small buttons are ideal for other cell phone users. One button size minimizes manufacturing costs.
4. Formulate the contradictions in the format of the standard TRIZ contradictions.	We want to optimize the area of a stationary object and increase the flexibility and versatility of the object.
5. Use the TRIZ matrix to identify standard recommended solutions.	The TRIZ matrix proposes that the principle of dynamics be investigated.
6. Translate the standard recommended solutions to possible solutions for your problem by understanding the resources available in your system.	The solution is the touch screen, or no buttons. The touch screen allows the user to customize button configuration through software. By eliminating buttons, and creating a dynamic input mechanism we have solved the problem.

What is preventing you from achieving the ideal state? Then, we want to understand the constraints in terms of contradictions. Maybe we want strong *and* light, fast *and* complex, small *and* easily accessible, or like in the cell phone example in Table 4.8 a small phone and accessible buttons. Next, we want to frame these contradictions in terms of the standard set of contradictions in the TRIZ matrix. There may be more than one set of contradictions requiring solutions, but we work in pairs. We can investigate many pair-wise contradictions during the TRIZ process. Using pairs of contradictions, we then go to the TRIZ matrix and study the design principles recommended for our sets of contradictions. There may be multiple principles recommended for each contradiction. These principles are intended to spark ideas to consider on your specific problem. The last step is to translate these design principles into possible solutions. At this point, we think the more possible solutions the better. We'll go through a process of weighing all possible solutions to choose the best amongst many.

This is a very brief overview of the TRIZ method. Please seek guidance and further training if TRIZ is a new tool for your organization. Later in the book, we will go through some additional examples of how TRIZ can be used in a SOFAIR project. This method is not only useful for physical systems and devices; it can be used on design innovative solutions to social systems.

4.4.3 7-Ways

The 7-ways is a tool used in the Lean production method of 3P (Coletta, 2012). It is a simple, but powerful tool. The concept is to find at least seven different ways to solve a problem, and this technique usually involving two-dimensional or three-dimensional simulation. It is an interesting phenomenon; finding three or even five different ways of solving a problem seems to be easy; finding seven different ways to solve a problem can be tough to do. It is in the last few reaches for new solutions that innovation happens.

Table 4.9 shows a simple example of a nutrition delivery system, how we eat our food. The first four or five are easy. The last three or four are difficult and require us to think in different ways. Creating a new fork or new spoon may not lead us to innovation. However, in this example, if we begin to think about intravenous nutrition delivery

Table 4.9 Nutrition Delivery Systems: 7-Ways

1. Spoon
2. Fork
3. Chopsticks
4. Food as container: tortilla, bread, bun, etc.
5. Food as liquid only: cup
6. Pills
7. Intravenous

systems we could conceive of nutritional skin patches or supplement injections. These ideas are far outside the norm of what we think about when we take a meal. It is the outside the envelope ideas that become the source of innovation.

To use the 7-ways, start by stating the ideal state. Just like with TRIZ, we don't want to describe a solution. We want to describe the ideal final result (IFR). Then, start brainstorming different solutions to achieve the ideal state. This is often done in groups. With large groups, it can be interesting to send subgroups to separate rooms to create independent 7-ways. And then bring the subgroups together, share everyone's 7-ways, and then collaborate as a larger group to find seven more ways, coalescing the independent ideas, into more and more innovative solutions. The 7-ways is an easy technique to influence teams to reach for innovative solutions.

4.4.4 Improve

As a continual improvement method, SOFAIR often results in incremental improvement. There are many benefits to incremental improvement. Minor process changes carry minor risks associated with change. Minor changes are easier for those operating in the process to accept and to learn. Minor changes are more easily reversed if new issues arise. Minor changes typically require fewer resources for implementation. Many minor changes over time result in major changes.

There are several steps to achieving minor process changes. The first step is to generate many possible solutions and choose the best amongst many. Then, before we permanently change our process we want to ensure that our knowledge about the relationship between process inputs and outcomes is valid in real-world application. We want to work out any kinks in our process changes before we affect the whole value stream with the changes. Lastly, we want to carefully plan and execute the change. This requires the use of project management tools. We want the implementation process to be as smooth as possible. We want to plan for effective communication and training. We want to make sure the period of change is inclusive and considerate of stakeholder needs.

4.4.5 *Selecting Solutions*

Thus far in the innovate and improve phase of the SOFAIR project we have used TRIZ, 7-ways and possibly other innovation tools to try to assemble as many possible solutions to our problems as possible. This is divergent thinking. Now we need to introduce some convergent thinking. We need a way to compare all of the possible solutions in order to choose and implement the best solution. One simple tool for accomplishing this is the prioritization matrix.

The first step in selecting solutions is to determine what defines the best solution. We need selection criteria. Table 4.10 shows an example solution matrix for a CISR problem. In this case, our team used the 7-ways method to find different solutions to improve access to healthy, nutritious food by young children in the community. The team has determined five criteria of importance: nutrition, quantity of food, frequency of access to food, ensuring that a broad range of children at different ages have access to the food, the frequency of access, and if the child's caretaker also has access to the food. However, these criteria are not of equal importance. In our prioritization matrix, we use a weighting scale using only the numbers 1, 3, and 9. The number 1 is used for low importance, 3 is mid-range importance, and a 9 indicates that this criterion has the highest importance. In this example, nutrition and age of access have the highest priority of importance. Then, each solution is rated for each criterion. In the rating scale, the numbers 1, 3, and 9 are also the only numbers allowed. This prevents the team from getting too involved in trying to determine the difference between a rating of 6 and a rating of 7, for example. It makes the rating process more efficient. And, it keeps the team from rating everything a "5," which would result in no prioritization amongst solutions. The rating is multiplied by the criteria weight, and then the total score for every solution is summed. In Table 4.10, we see that the

Table 4.10 Prioritization Matrix—Improving Child Access to Free Meals

POTENTIAL SOLUTIONS	SOLUTION CRITERIA					TOTAL SOLUTION SCORE
	NUTRITION	QUANTITY OF FOOD	FREQUENCY OF ACCESS	AGE OF ACCESS	CARETAKER ALSO HAS ACCESS	
Criteria Weight	9	3	3	9	1	
State provided vouchers for school breakfast and lunch	9(81)	3(9)	1(3)	1(9)	1(1)	130
Charity kitchens promoted in home neighborhoods	9(81)	9(27)	9(27)	9(81)	9(9)	225
Vouchers provided for grocery purchase in home neighborhood	3(27)	9(27)	9(27)	9(81)	9(9)	171
Charity food pantries promoted in home neighborhood	3(27)	9(27)	9(27)	9(81)	9(9)	171
Community gardens zoned and maintained	3(27)	3(9)	1(3)	9(81)	9(9)	129
State audit of home environment with placement of malnourished children to state institution	3(27)	9(27)	9(27)	3(27)	1(1)	109
No state influence in child access	1(9)	1(3)	1(3)	1(9)	1(1)	52

preferred solution is to promote charity kitchens in neighborhoods. The prioritization matrix score of this solution is much greater than any of the other solutions.

Another benefit of the prioritization matrix is the use of the scoring to create new, hybrid solutions. In our example, the solution of placing food pantries in the neighborhood scored high, but not as high as kitchens. The pantry solution rated lower on the criteria of nutrition. The concern is that caretakers without a proper understanding of nutrition may select lower nutrition food from the pantry for home preparation. Analyzing how this solution scores lower can give ideas on further improvements for this solution. What if some high nutrition pantry items were offered for take-home at the neighborhood kitchens? This would further improve the neighborhood kitchen idea.

Choosing effective criteria is a critical part of the solution selection and prioritization matrix. The discussion during criteria and criteria weight development is a great teamwork opportunity. This is also a good time to go back to stakeholder input information. The prioritization matrix can advise the team on the best solution amongst many possible solutions. And it can be used to analyze the pros and cons of different solution to innovate for even better hybrid solutions.

4.4.6 Improvement Implementation

The last step in the innovate and improve phase is to implement the improvement. Process change management and project management techniques are applied now. Process change management tools are primarily centered on communication techniques. We will briefly discuss an action plan as a project management tool. The act of implementing process change involves makings sure that everyone affected by the change knows about the change before it happens, allocating time and resources to make the change, and then training and controlling the adoption of the change. For large, complex process changes we advise starting the change with a trial part of the process. Affecting the change in a single department, or single location, or single work group, in order to work out the difficulties and risks of the change, can be one way of reducing problems associated with the process of changing. When implementing improvements, communicating and allocating time and resources are critical factors to the success of leading the process change.

Communication starts with a plan; it doesn't just happen. And different stakeholders will need different types of communication. Communication can be synchronous (when we are all in the same room or phone call at the same time) or asynchronous, such as a website or email. Some stakeholders will be eager adopters of the change, some will be resistant. Some stakeholders will be integrally involved in the change; other will only be involved in the periphery. Figure 4.18 demonstrates some communication strategies for different scenarios. In all cases, all persons affected by the change should be communicated with before, during, and after the process change. The purpose of communication is for the person initiating the change to become aware of new issues that may need to be resolved during the change and for the person affected by the change to effectively change their behavior in order to achieve the new level of performance. Communicating change is a two-way street.

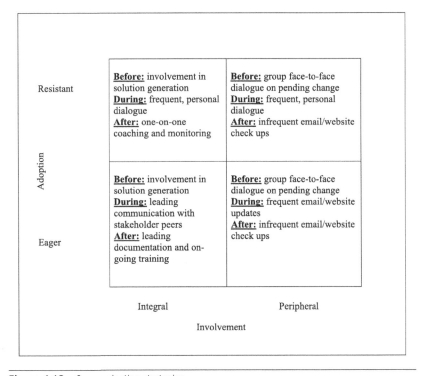

Figure 4.18 Communication strategies.

There are a variety of project management tools that can be used. Which tool you use depends on how simple or complex the project is. For most SOFAIR projects a simple action plan will suffice. Table 4.11 show a simple action plan, showing responsibility lanes, for a process change. This type of action plan shows simultaneous action in different departments laid out in "swim lanes." It also allows for a single person to be responsible for the cross-functional organization during each phase of implementation. In most large organizations, there are professional project managers who can help with deploying project management tools for complex process changes, especially when the change involves equipment or software installations. We encourage the SOFAIR project leaders to find additional sources for project management tools and techniques. The Project Management Institute (n.d.) is a resource for this type of information. There is so much written elsewhere on project management we will not belabor the tools here. However, this is not to indicate the unimportance of effective project management. Allocating resources and coordinating effort across time is a critical part of the SOFAIR project.

After all the hard work in the stakeholders and subjects phase, the objective phase, the function and focus phase, and the analyze phase,

Table 4.11   Implementation Action Plan

PURCHASING	ACCOUNTING	PROCESSING	PACKAGING	ACTION LED BY AND DUE DATE
Enter new "non-GMO" requirement into order specification	Get purchase order placed for non-GMO quality inspections			Mary May 13
Enter qualified suppliers into database				Mary June 15
Add receiving inspection criteria for non-GMO labeling and supplier approval		Check inventory releases for non-GMO status and labeling		Bob June 30
		Process only on non-GMO allocated equipment	Ensure finished product labeling has non-GMO disclaimer	Jane July 5

it is now time to pay attention to making process changes. Without proper communication to those who are being affected, and without the proper allocation of resources, all of the hard work in the previous phases can be negated. It all comes down to effecting the process change. Without change there is no improvement. Implementing the process change is a critical part of a successful SOFAIR project. Consider a trial of the process change on a small scale first. Take your time to plan your communication and manage your resources.

4.5 Report and Repeat Phase

The primary principles of social responsibility include accountability and transparency. Accountability is the concept that an organization is responsible for its impacts on all stakeholders. It is the responsibility aspect of social responsibility. Transparency means that the decisions and actions that lead to impacts should be shared with stakeholders. Accountability and transparency are linked; transparency is the antecedent and accountability is the consequence. The use of reporting is one way of acting with accountability and transparency. Sharing outcomes and impacts and the strategies and decisions associated with outcomes and impacts in a public forum allow for transparency; information in the hands of stakeholders can then lead to accountability. Therefore, reporting is a key part of continual improvement for social responsibility, and thus a phase of the SOFAIR method. In this chapter, we will discuss different ways to report, both internally and externally.

As a continual improvement methodology, we also include the concept of repeating as a part of the SOFAIR method. The end of a SOFAIR project is marked by the beginning of the next round of improvement. We are never finished; there will always be increasing stakeholder expectations and new opportunities for performance improvement. The way to approach improvement opportunity should not be accidental or haphazard. Social responsibility improvement should be in service to the organization's success strategy; the ultimate goal is 1500+ years of sustainable business success. Therefore, improvement targets and objectives should be aligned to business strategy. In the following pages we will discuss techniques of reporting and repeating the SOFAIR method that achieves excellent CISR performance while minimizing wasted effort and distractions.

4.5.1 Reporting as Strategy

Reports are periodic snap shots of performance. Ideally, reports contain both qualitative information and quantitative information. Information about decisions, activities, and impacts should be included. We recommend approaching reports with a system's perspective. The overall business environment should be discussed in the report. This will set the stage for the intentions that lead to your decisions. The organization operates within the business environment; how the organization does this is its context. What is the organization's mission in consideration of the larger business environment? Vision? Goals? Objectives? This is an opportunity to tell the organization's story. Having the stakeholder understand the organization's narrative of purpose will also set the stage for the intentions that lead to decisions. The business environment and organizational context are qualitative data. This is textual reporting. Case studies, examples, and even personal stories of leadership perspectives can be effective parts of the report. Next, the organization's decisions, activities, and impacts should be shared. This information is typically more quantitative. Input and output metrics can be considered. Comparisons such as year-over-year performance, comparisons to competitors or similar organizations, industry statistics, performance indicators, and monitoring metrics can be covered here. Stakeholders are most interested in impacts. The point is not to be exhaustive in metric reporting, rather to be intentional in metric reporting according to stakeholder interests.

Think of a report as a part of stakeholder dialogue. Yes, it is a monologue, one-way communication between the organization and the world. But if published with the expectation that the report is merely the start of a conversation with stakeholders, feedback from the report information will be better integrated. This keeps the report from being something extra beyond stakeholder dialogue. If done adeptly, the report can be a part of stakeholder dialogue. It also helps to keep the report brief. The goal is not to make as big a report as possible; the goal is be thoughtful and intentional in providing the necessary information to demonstrate accountability and transparency to only the organization's stakeholders.

4.5.2 *Internal Reports*

There are many internal stakeholders in all organizations. Employees are usually the largest body of internal stakeholders. Management and leadership are stakeholders. Departments may be stakeholders of each other. Contractors and service providers can be acting as internal stakeholders. In some very short supply chains, even customers and suppliers can act as internal stakeholders. For example, in a university, students are stakeholders. Although not employees of the school, students are so integral to the value stream that they can be considered internal stakeholders, not customers. Internal reports serve these internal stakeholders.

Most external reports are published on an infrequent basis, such as annually. Internal reports should be published much more frequently. A benefit of internal reporting is to motivate internal stakeholders to work to improve the performance of the organization. Frequent internal reporting helps to engage organizational members to prevent and correct social responsibility performance. Monthly reports are more appropriate than annual reports for this purpose. Table 4.12 is an example of a social responsibility scorecard that could be used to inform internal stakeholders about concerns, trends, and performance. Note that all of the elements of effective reporting are present: business environment, organizational context, decisions, activity, and impact. In some organizations, these types of scorecards are published electronically in cafeterias, hallways, or other information centers. These reports can be pushed via email or intranet websites. Whatever the medium, they should be accessible to all internal stakeholders.

We should provide a word of caution on information that is intended for internal stakeholders that might leak to external stakeholders. Internal stakeholder reports should be comprised of information that if leaked external would not cause a concern for external stakeholders. However, because of the nascent aspect of the information, the accuracy and validity of the information in internal reports is usually not as rigorous as the same information may be in an external report. Information in external reports is typically thoroughly audited and tested prior to publication. This is not necessarily the case for internal reports. For this reason, it is best to attempt to keep internal

Table 4.12 Internal Reporting: A Special Responsibility Scorecard

MAY 2015: ACME MANUFACTURING, PACKAGING DEPARTMENT—SOCIAL RESPONSIBILITY SCORECARD			
ACME STATE OF THE BUSINESS	KEY DECISIONS	PACKAGING DEPARTMENT INDICATORS	PACKAGING DEPARTMENT INTERNAL METRICS
Acme has seen its three straight month in increasing orders. We are scheduled to package 14,740 units this month. • 52% of deliveries to Asia • 15% of deliveries to Europe • 18% of deliveries to North America • 15% of deliveries to South America The assembly department has a 98% on-time-delivery performance. The cardboard supplier has a 100% on-time-delivery. However, the bubble wrap supplier is behind schedule with a 65% on-time-delivery. Packaging operator flex schedules will be in effect for the month of May due to this disruption.	Tickets for the company picnic are now on sale. Earlier purchases help us better plan the food and space required. If you plan to attend, please purchase your tickets now. Help us prevent food waste! With the on-going schedule disruptions caused by our bubble wrap supplier, and the negative environmental impact of plastic bubble wrap we are initiating a SOFAIR project to minimize or eliminate the use of bubble wrap in our packaging. If you would like to participate in this project, contact Carol in the Quality Department.	Governance 1. Participation in CISR. Human Rights 2. Suppliers screened for human rights compliance. Labor Practices 3. Safe behavior compliance. Fair Operating Practices 4. Promotion of local sourcing. Consumer Protection 5. Packaging design for safe transport and opening. Environmental Protection 6. Minimum and recyclable packaging materials. Community Development 7. Employees' family involvement in sustainability efforts.	Governance 1. New SOFAIR project started on minimizing bubble wrap. Human Rights 2. 96% of suppliers screened for human rights. Jerry is auditing 2 more suppliers in June. Labor Practices 3. May TCIR: 0.5, YTD for packaging department TCIR: 1.2. Fair Operating Practices 4. May: 42% of supplies are sourced local by pounds purchased. Consumer Protection 5. New tape testing is on-going in R&D with zip removal. Testing to be completed in July. Environmental Protection 6. May: 75,922 pounds of recyclable packaging material shipped. YoY 7% reduction. Community Development 7. Family picnic planned for July. Children's "sustainability science fair" email sent to all employees.

reporting as internal as possible. Disclaimers on the reliability of the data should be made, if necessary. In an ideal world, internal reports could be external reports, with perpetually available data of perfect accuracy. However, for most organizations this is not practical.

A good internal report is as brief as possible. As one aspect of stakeholder dialogue, internal reports should only focus on those impacts of concern to internal stakeholders. A good internal report will provide information that is actionable. A good internal report will advise internal stakeholders on how they can affect social responsibility performance improvement. Internal reports can motivate internal stakeholders to change their behavior. Internal reports should advise internal stakeholders on how to answer these questions concerning social responsibility: On what should I be concerned? And, how can I help?

4.5.3 External Reports

There are two varieties of external reports: sustainability reports and integrated reports. Many publicly traded companies are required by their stock exchanges to publish annual public reports. Many organizations choose to integrate their sustainability and social responsibility performance within these, or other shareholder-oriented reports. These are referred to as integrated reports. Other organizations choose to publish a separate sustainability or social responsibility report. There are many reasons for doing this. There might be information of interest to nonshareholder stakeholders that is not of interest to shareholders, and thus a separate sustainability report is desired. The organization may choose to publish sustainability performance information more, or less, frequently than shareholder-oriented reports. Or, the organization may choose to highlight its sustainability efforts and use a separate report to reach an audience beyond that of shareholders.

The purpose of all reports, shareholder-oriented, integrated, or sustainability is to begin a dialogue with stakeholders. Choosing which type of report is appropriate for your organization depends on your stakeholder's interests, your impacts, and other aspects of stakeholder dialogue. Table 4.13 shows a small list of some available external report forms. There are report formats that apply to many different subjects of both sustainability and social responsibility; these are

Table 4.13   External Reporting Forms

REPORTING STANDARD OR FRAMEWORK PUBLISHER: FRAMEWORK TITLE	INDUSTRY OR SPECIALIZATION
• The Global Reporting Initiative: GRI G4	Universal
• International Integrated Reporting Council: <IR> Framework	
• Global Initiative for Sustainability Ratings: Center of Ratings Excellence (CORE) Program	
• United Nations Global Compact: Communication on Progress	
• Social Accountability International: SA8000 Standard	
• Sustainability Accounting Standards Board: SASB Conceptual Framework and Industry Briefs	Accounting
• Dow Jones Sustainability Index: Corporate Sustainability Assessment	Stock market-related
• CDP (previously known as Carbon Disclosure Project): Guidance for Responding Organizations	Environment, only
• United Kingdom Department for Environment, Food and Rural Affairs: Environmental Reporting Guideline	
• Electronic Industry Citizenship Coalition: Code of Conduct and Assessment	Electronics industry

listed as universal reports. There are other reports that are focused on certain industries or professions. And there are stock-market-oriented reporting formats that have been developed for the socially responsible investor. Your organization may choose one of these forms or create its own. We only advise that your reporting is aligned with your stakeholder dialogue intentions.

External reporting is the ultimate in transparency to all possible stakeholders. However, even with the potential for very broad distribution, there should still be brevity in reports. The reports should align with business strategy. It is not necessary to report on every possible indicator and metric in a reporting framework. The organization should evaluate its decisions, activities, and impacts and align report metrics only on those subjects that are important to stakeholders and the sustainability of the organization. For example, it may not be material for a small nonprofit provider of services for elderly clients in Canada to report on its water footprint or have a mechanism for human rights grievances. However, reporting on employee diversity and equal opportunity or ethics and code of conduct may be very important to stakeholder dialogue, and thus should be reported.

Whether the organization chooses internal or external reporting, integrated or separate sustainability reporting is a decision that should be carefully evaluated. However, this decision can be revisited at any time. The important factor is deciding what reporting format best initiates productive stakeholder dialogue. And, assuming the organization is dedicated to CISR and SOFAIR, continual performance improvement in important reporting metrics can be demonstrating, leading to continued positive stakeholder engagement.

4.5.4 Repeat

As a continual improvement methodology, the end of one SOFAIR project leads to the beginning of the next project. This is the repeat phase of every project; as we exit the project we choose the next opportunity for improvement. The continual improvement approach does two things to accelerate social responsibility performance improvement: first, since each project is scoped for success, success builds confidence and motivation; secondly, since we end by starting the next round of improvement we are always improving. After the completion of a few projects, it can be amazing to see how fast the organization has shifted perceptions on social responsibility performance and gained a sense of mastery.

4.5.5 Continual Improvement

We take our continual improvement methods from the quality profession. These methods are well-worn in this sphere. Most large, mature organizations already have some form of continual improvement program in place, such as Six Sigma or Lean production. We strongly encourage that your CISR efforts mimic these other programs, at worst; we recommend that CISR is thoroughly embedded in your continual improvement programs, at best. Use the resources and structure that you already have, if possible.

The repeat phase uses many tools. Continual improvement methods are founded on the plan–do–study–act cycle, referred to as Deming's PDSA cycle (n.d.). The SOFAIR method can be conceived of in the PDSA cycle (see Figure 4.19). How much time you spend in plan versus do, or check versus act varies from project to project. The main point is that it is a never ending cycle. The minute success is achieved in one aspect improvement is sought in another. Another important

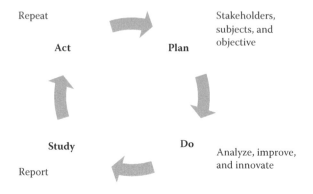

Figure 4.19 Plan—do—study—act.

lesson from W. Edwards Deming, associated with the concept of continual improvement is "constancy of purpose" (Deming's Fourteen Points for Management, n.d.). Remember, constancy of purpose is the first of Deming's 14 points for management (see Table 1.1). A "flavor of the month" approach will not result in high performance. The type of performance improvement needed for both quality and social responsibility happens incrementally over very long periods of time. Thus, constancy of purpose, or fortitude and commitment are needed by management to stick to the program over time. Planning for, and committing to, five or more years of SOFAIR projects would be an example of constancy of purpose. We are not saying that it takes that long to see results; we are saying that unless a long-term commitment to the method is deployed, the fundamental cultural changes to sustain the gains will not happen.

If your organization does not have a continual improvement program, and you will be initiating this type of activity as a part of your CISR initiative, we recommend starting continual improvement with Hoshin Kanri. Hoshin Kanri was discussed in the objective phase (see Figure 4.5). It is a method to deploy strategy throughout the organization. And this technique operates cyclically, as a continual improvement process. Hoshin Kanri deploys key performance indicators and improvement targets. In the spirit of continual improvement, once one level of performance is achieved the next level of improvement is targeted. Using Hoshin Kanri to manage your continual improvement efforts allows for the next level of performance improvement to be easily targeted.

4.5.6 Choosing the Next Opportunity

The last activity in a SOFAIR project is to choose the next project. Through the project closure information and data have been gathered that can help with setting the next priority. Stakeholder dialogue has resulted in a body of information that can be used to determine the next opportunity. Risk assessments can point to the next level of concerns. And life cycle analyses can reveal potential areas of improvement. Using project closure to propel energy and focus on the next project builds a strong continual improvement foundation. Don't leave one project without identifying the next.

Review the stakeholder and subjects phase of the project. Which subjects were excluded from the focus of this project? Is there an opportunity for improvement on the same process and scope, but with different subjects? In this case, you may choose to keep the existing team together to tackle the additional subjects. Was there information from the stakeholder dialogue that was excluded from the study? Were there stakeholder groups that were minimized from focus? Should there be a new project chartered on the same process for these additional stakeholder interests? Review the analyze phase for additional opportunities. Calculate your new RPN on your FMEA. What are the next five highest RPNs? Are these important risks to consider in the next project? Take a look at the life cycle analysis for this project. Are there life cycle phases that are in need of further improvement? If so, you may be able to launch the next project skipping straight to the analyze phase, just reviewing the "S," "O," and "F" phases of the previous project in consideration of the new focus.

Throughout the SOFAIR method, we have intentionally worked to narrow our focus in every phase. Developing this focus is important for making meaningful and manageable improvement. In the report and repeat phase, we should go back and review all the topics and issues that were jettisoned to bring us to the focus of improvement achieved in this project. The review of this information usually leads to an easy identification of the next, most important, priority of improvement.

4.5.7 Building a CISR Culture

Closing a SOFAIR project in an organization with a culture that has continual improvement for social responsibility embedded helps to sustain the gains achieved in the project. As summarized in Table 2.2, there are five factors of success in a CI program are organizational support, clear roles and responsibilities, a rigorous methodology, linkage to the business strategy, and measurable outcomes. Top leadership must be willing to commit resources and support for CISR. There should be well-trained professionals dedicated to the CISR effort. A rigorous methodology, like SOFAIR that prevents green-washing and spin ensures that valid improvements are achieved. All of this CISR effort is of no use if the organization is not sustainably successful; CISR efforts must be linked to business success strategy. And outcomes must be measurable to monitor sustained success.

Translating repeated success associated with completed SOFAIR projects into culture change involves some action. Table 4.14 identifies some things that the organization can do over time to help discrete SOFAIR projects yield broad organizational change. Involve as much of the organization in SOFAIR as possible, deeply and widely. Recognize, reward, and communicate SOFAIR success. Involve external stakeholders to help spread the methodology. Recognize that every SOFAIR project brings the organization closer and closer to a CISR culture.

Table 4.14 Turning Project Accomplishment into Culture Change

- Involve all levels of organizational members in SOFAIR projects
  - From executives to front line employees
- Involve all functions and departments in SOFAIR projects
  - Mix teams with diverse members from different functions
  - Ensure that all functions are leading projects
- Treat SOFAIR project leadership training as a professional development privilege
  - Recognize and reward SOFAIR leadership
- Connect social responsibility performance outcomes to bonus programs
  - Promote measureable outcomes and valid improvement results
- Involve top level CISR program leadership in business strategy development
  - Challenge business strategy to be thinking about sustainability
- Broadcast SOFAIR project success across the organization
  - Promote talent commitment to the organization
  - Help organizational members understand what is possible
- Involve suppliers, customers, and community members in SOFAIR projects where practical
  - Begin to set different expectations with these stakeholders
  - Spread CISR and SOFAIR beyond the organization

Through the repeated use of SOFAIR, embedded in a CISR program, eventually leads to a CISR culture. Depending on the size of the organization and the extant culture, this may take years or even decades. Constancy of purpose and broad inclusion and engagement across the organization in CISR helps to initiate this culture change. The end goal is not just projects or programs. The end goal is an organization that has core beliefs, values, and behaviors that result in sustainability and social responsibility. The end goal is a CISR culture.

5

Examples of SOFAIR
in Action

We've spent a lot of time discussing methodology and methods. We've discussed roles and responsibilities and how CISR®* can be deployed as a program in your organization. We talked about the phases and tools and techniques that can be used in a SOFAIR project. We've demonstrated how most of this approach is a spin-off from the Six Sigma methodology. Building on such a proven methodology means that we can accelerate the achievement of results. We don't have to start from the beginning. We can launch from what has already been learned.

In this chapter, we will be sharing some real-world examples with you. This will allow you to see a whole project from start to finish. You will be able to see how some of the tools are used. You will be able to see how each project differs from another just a little bit. You will begin to get an appreciation for the different types of issues that can be improved with SOFAIR.

Then, in the next chapter, we have lists of actions that can be taken, immediately, by different members in the organization. We want you to leave this book motivated and ready to take action. Talking about social responsibility, strategizing about social responsibility, planning, and even complaining about social responsibility doesn't get us to prosperity for the next 1500 years. We need to step up to our responsibilities today. We have identified many different small, but impactful, and readily taken actions that everyone in the organization can take today to make a difference.

---

* CISR® (sounds like scissor) is a registered trademark and can be used with permission for non-commercial use. Contact SherpaBCorp.com for permission.

5.1 How to Get Started

The following examples will help you understand how to take a first step. CISR and SOFAIR are not theories; performance improvement requires action. But in large, complex organizations, how to take the first step might not be readily apparent. Political influences might be needed. Resources need to be allocated. Leadership commitment needs to be visible. Behaviors will change. Possibly tense stakeholder dialogue needs to take place. It is understandable if, at this point in time, you are ready to start moving forward with CISR, but you don't know where to begin.

One of the benefits of the CISR and SOFAIR approach is that it can be launched with a small scope. One business, department, or group can start a SOFAIR project without the full complement of the organization. Sometimes, just proving what can be done at the grass-roots level can influence top leadership. For example, if you work in a small shipping department in a large organization, start a SOFAIR project on the sourcing of your shipping supplies. Or start a SOFAIR project on your warehouse cleaning materials and process. You can conduct your first SOFAIR project on a very small scale.

Another way to start small is to begin to integrate some SOFAIR projects in a mature continuous improvement organization. If you work in a large, complex organization, you probably have a quality assurance or continuous improvement department. Approach one of the performance improvement leaders about getting a SOFAIR project introduced into the organization's continuous improvement project portfolio. The first project doesn't even have to be about social responsibility. Think back to our example with the tomatoes. The project was launched due to a customer satisfaction issue. But, by using the SOFAIR method we were able to delight all stakeholders, not just customers.

If you are a social responsibility or sustainability leader in a large, complex organization, one small way to start is by hiring a single SOFAIR project leader. Even with a limited budget, we would encourage corporate social responsibility leaders (often termed CSR) in large organization to hire problem solvers, rather than consultants, report writers, or public relations professionals. Hire just one project leader trained in SOFAIR. This person will be able to lead 4–6 projects per year. And then, let the organizational pull for projects pace the hiring of additional project leaders.

It is critically important, even when starting small, that your CISR efforts are aligned to business strategy. One project in the shipping department, or one SOFAIR project added to the continual improvement docket, or one full-time SOFAIR project leader working for the CSR executive must be working on performance improvement efforts that are meaningful to the business. The SOFAIR method is to be applied to *all* business problems, not just social responsibility problems. The examples that follow will give you an appreciation for the different types of issues that can be resolved by SOFAIR. In all cases, the business must succeed through CISR. The point of CISR is to have a successful business 1500 years from now. Business strategy must be an integral part of the CISR and SOFAIR launch.

Our last point, before sharing some examples with you, is to encourage you to just do something. The next chapter has ready actions for organizational leaders, project leaders, communicators, and team members to take, now, to move social responsibility performance improvement forward. The end goal of CISR is culture change. We will know we are "there" when everyone in the organization thinks and behaves in a way that is sustainable for at least the next 1500 years. The spark for this culture change is initiated by taking action and seeing results. Do something. Today.

5.2 SOFAIR Example: Healthcare

This SOFAIR example applies to the healthcare industry. Hospitals, pharmaceutical companies, home healthcare providers, doctors, nurses, dentists, physical therapists, or anyone providing products or services to deliver sustainable human health can benefit from reading this example. In this scenario, we have a service provider of kidney dialysis in a small community whose members are predominately members of a Native American tribe. Unfortunately, in the United States, there is an epidemic of diabetes amongst Native American people. Diabetes can eventually lead to renal problems, and there is intense growth with new dialysis treatment centers in Native American communities. In this example, we will play the role of "Dia'mond," a dialysis service provider.

We are projecting a 27% growth in the coming year, which will mean opening five new centers in the southwestern United States. Dia'mond is taking a 1500 year view of their expansion. We want to become a productive and integral part of the new communities in which they are expanding. We want to be a healthcare provider, not a sick care provider. Follow the SOFAIR process and we will see Dia'mond makes expansion decisions in these communities that result in healthier options.

5.2.1 Stakeholders and Subjects

We start the first step of the SOFAIR project by engaging with stakeholders. We have selected three of the communities that are targeted for dialysis site expansion. Before we build the new facilities, we want to understand the stakeholders of the new sites and understand how the seven subjects of social responsibility can be considered. We want our expansion to be socially responsible.

We start with the SIPOS, shown in Figure 5.1. In this process, we have a seemingly never-ending loop. Once renal failure has set in, we can anticipate that the patient will be dependent upon dialysis for the

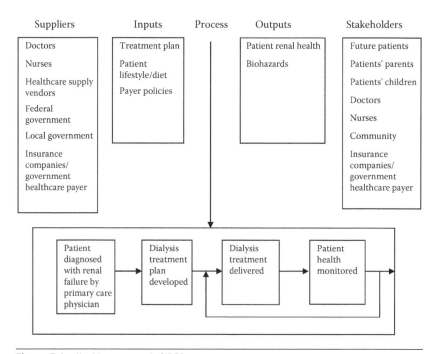

Figure 5.1 Healthcare example SIPOS.

rest of his or her life. This is a process to be avoided. We also see that our stakeholders include our patients, their extended families, and the community at large. Also, those involved in the healthcare industry are also stakeholders: doctors, nurses, insurance payers, and others.

We start our stakeholder engagement with a focus group of community elders. They have extended the invitation to spend a whole day with tribal elders understanding how diabetes and dialysis have affected the community. We find how devastating the diagnosis of renal failure is to the patient, family, and community. It is seen as a death sentence since most patients are older and renal failure marks the beginning of the end of the health crisis of diabetes. We do a full-day focus event in all three of the expansion-targeted communities.

We also spend time with doctors and nurses at a few of the existing dialysis sites. Engaged with these stakeholders, we find that missed appointments and late arrival to appointments are an issue. There is a higher incidence of hospitalization in native communities as a consequence of missed appointments. This is a result of difficulties with transportation to the dialysis clinic. Many older patients are reliant on children or grandchildren for transportation to dialysis appointments. Homes are very rural and far from the treatment site.

We spend time with insurance companies and government providers of healthcare payments. We find that the cost of dialysis is one of the largest healthcare costs for these patients. The insurance companies are very eager to work with us to do anything to improve or even prevent patients from entering into the revolving door of dialysis.

Lastly, we review all seven social responsibility subjects. Although we are eager to start on our CISR project, this is an important last step of data gathering to ensure that we thoroughly understand the full scale of the problem before we start setting objectives for the project. Figure 5.2 shows our assessment of the seven subjects. Here we have a potential dilemma. The organizational governance and community involvement and development aspirations are potentially in conflict. We are in the business of providing dialysis. If we work with the community to prevent diabetes and renal failure, then we minimize our profit. Yet, we don't want to promote diabetes! We will want to keep this conflict in our sights during our SOFAIR project. If we can find solutions that both maximize Dia'mond's profit and minimize diabetes, this will be a social responsibility "home run."

---

1. Organizational Governance
    a. We should maximize Dia'mond's revenue while protecting the long-term health of the dialysis patients. We should not maximize our profit at the expense of our patients' health. Governance structures to measure patient health should be put in place.
2. Human Rights
    a. We should ensure that all healthcare workers have legal work status.
3. Labor Practices
    a. We should ensure that all healthcare workers have appropriate working conditions, compensation, and work hours.
4. Environment
    a. We should ensure that all biohazards are appropriately disposed.
    b. We should ensure that the sites are maintained for energy efficiency.
5. Fair Operating Practices
    a. We should ensure that we compete fairly with other dialysis providers.
6. Consumer Issues
    a. We should ensure that all community members have access to our services.
7. Community Involvement and Development
    a. We should work with the community to minimize or eliminate diabetes and renal failure.

Figure 5.2 Healthcare example evaluation of the seven subjects.

5.2.2 Objective

Now that we understand our problem better, we can begin to formulate the objective of the project and our desires for continual improvement for social responsibility. Our example objective for this problem is shown below.

Dia'mond seeks to expand operations and to be a responsible provider of dialysis in rural communities. We want to promote health within the communities we serve. We want renal failure and dialysis to be avoided; but when needed, we want to treat our patients and their families with dignity, compassion, and quality care.

This is a complex problem. As noted previously, there are potentially contradictory goals within this objective. How do we simultaneously improve community health and grow Dia'mond's business success? How do we balance people *and* profit? This is a difficult objective.

5.2.3 Function and Focus

Now we know what everyone involved in the decision needs and wants. We know what stakeholders want and we know what the organizational objective is. We understand that there are many different subjects to consider with our decisions. We have a difficult objective; we will need to focus our efforts. Taking a large complex problem and breaking it into manageable elements will work best. Then, once one small issue is solved, we will use a continual improvement process to tackle the next issue, and then the next issue, and then the next. This keeps us from becoming overwhelmed with such an uncertain, complex, and ambiguous scenario. Focusing on a small, manageable piece of the problem frees us to take immediate action toward improvement.

We use the value function diagram to start to brainstorm opportunities for focusing our problem solving. Figure 5.3 shows our value function diagram, with some failure modes identified. This work was completed with a cross-functional team. We chose one of our existing rural sites and gathered together three patients, one newly diagnosed and two who have been receiving dialysis for over a year. We pulled in two nurses and a doctor. And we had a tribal elder. Together, the team of seven stakeholders came up with these potential social responsibility issues. The team then voted to focus on the first process step, the diagnosis of renal failure. The earlier dialysis can begin, if needed, and the earlier warning signs are heeded, the better off our patients will be.

The communities targeted for dialysis center expansion are extremely rural. Access to primary care doctors, much less treatment centers, is difficult. It is not uncommon for people to drive 40 miles

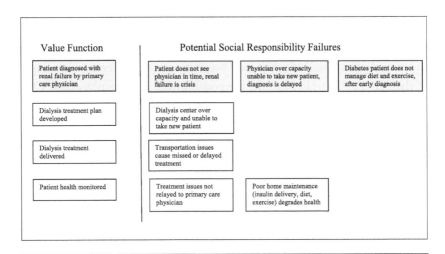

Figure 5.3   Healthcare example value function diagram.

or more on dirt roads to access a doctor, a grocery store, or a gas station. Health crises are probable if the progress of the diabetes is not very carefully managed. We have little hope of achieving our organizational objective of promoting health within our communities if we don't improve earlier access to healthcare and treatment. We should work hard to prevent the progression of the disease.

5.2.4 Analyze

Because our focus is on the diagnosis of renal failure, this is the only process step that will be carried forward for cause-and-effect analysis. We are not indicating that the treatment plan, the treatment delivery, and the patient monitoring are not important, nor will they be ignored for performance improvement. But we will not work on these process steps *now*. For now, we are focused on diagnosis. After we solve our problems and implement improvements to diagnosis, we will turn our attention to the other process steps. It is critically important to focus in this way. With new systems of diagnosis, the subsequent process steps may change, stakeholder needs may change. Working on everything at once causes confusion and chaos; some solutions may work against, or negate, others. We want to take a slow and steady methodological approach to our opportunities.

In this example, we have used a fishbone diagram to better understand our cause-and-effect relationships. In this case, the effect is delayed diagnosis of renal failure. Again, with the same team of patients, healthcare providers, and tribal elder, we brainstorm the causes of disease escalation leading to delayed diagnosis. Figure 5.4 shows the results of our cause-and-effect analysis. We can begin to see the interconnectedness of the patient, his or her family, the community, and the physician. Our solutions will need to be holistic and complex to deal with these interactions. There are many factors that cause the disease to escalate, leading to a healthcare crisis.

5.2.5 Innovate and Improve

Now we know many of the causes of how patients arrive on Dia'mond's doorstep. And we are still faced with a dilemma. On the one hand, we want to be socially responsible and keep patients from needing

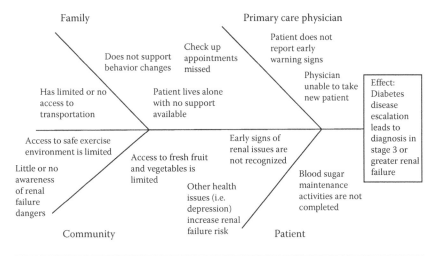

Figure 5.4   Healthcare example fishbone diagram.

life-saving dialysis. On the other hand, we are in the business of providing dialysis services. If we eliminate the need for our patients' dialysis requirements, do we put ourselves out of business? We have a conflict. However, there is an innovation tool for conflicting needs: TRIZ. We will use TRIZ to find an innovative solution to solve this conflict. We want a solution that simultaneously achieves the objective of promoting health in the communities that we serve *and* maximizing our business success, sustainably.

Our primary contradiction is to improve return on investment, or profit, associated with opening new treatment centers, while improving community member health by avoiding dialysis treatment. Our fishbone diagram resulted in a theme that leads to the conclusion that the earlier we can detect renal issues, the better we can avoid the need for dialysis. Using the TRIZ matrix (Solid Creativity, n.d.), we find two TRIZ contradictions that will be useful to study: quantity of substance (in this case, profit) and difficulty in detection (in this case, renal failure) (see Figure 5.5).

One of the solutions that the TRIZ matrix recommends for this contradiction is to think about local quality. This means that different parts of the product being designed have different, and local, functions. An example of local quality is a pencil and eraser being combined. One end of the pencil has the local quality of writing; the other end of the pencil has the local quality of erasing. Another example

Primary Conflict: Improve profit by decreasing the customers need for the service

TRIZ Conflict:

Quantity of a Substance

  Our desire is to increase our return on investment for new treatment centers

Difficulty in Detecting

  Our desire is to improve the ability to detect the warning signs of renal failure as early as

possible

TRIZ Solution:

Local Quality

Make each part of an object fulfill a different and useful function

Figure 5.5   TRIZ conflict and potential solutions.

of local quality is a cork screw with a bottle opener; one end opens a wine bottle; the other end opens a beer bottle. Each end has a local quality, unshared with the other end, yet on the same device.

This leads us to think about how we can create treatment centers that have different functions with "local quality." Our problem solving team comes up with a brilliant solution from the TRIZ suggestions. Why not include primary care, exercise, and food preparation/grocery services in the same building? From the fishbone diagram, we are reminded that transportation is a root cause of poor health. Access to nutritional food and safe exercise facilities are root causes of poor health. Early access to primary care physicians is a root cause to poor health. If Dia'mond expands their concept of a "treatment center" to a "health center," perhaps the goals of increasing community health and increasing profits can be achieved simultaneously. Dia'mond innovates their business concept and strategy. The new facilities will be expanded to include a clinic for primary care, an exercise gym with physical therapy and training, and a café with diabetic-diet friendly prepackaged frozen food, and frozen fruits and vegetables. In one visit, the patient can receive all the services needed to prevent degradation of health; Dia'mond becomes a responsible and respected key member of the community.

5.2.6 Report and Repeat

As an outcome of this project, originally focused only on early detection of renal issues, the entire business model of Dia'mond has changed from providing services after patients have developed healthcare crisis, to one of also providing preventive care services. The expansion of the business strategy has the potential to significantly increase revenue and profit for Dia'mond. The expansion of the business plan also has the potential to significantly improve the health of the community. Dia'mond now has plans to expand existing facilities in rural communities to provide primary physician care, physical training services, and food services. There will be new investments, potential difficulties with competitors in these communities (if they exist), and still potential issues with patient lifestyle and behavior changes. There are many details yet to be determined with this new solution. However, the social responsibility, and most importantly, the sustainability of Dia'mond as a successful business in the communities in which it operates are improved. Now, on to the next step in the process flow diagram.

5.3 SOFAIR Example: Manufacturing

Manufacturing, especially of consumer products, is a natural area for SOFAIR to be used since there are so many stakeholders throughout the product lifecycle. Our example is All-Belle Doll Company, founded in 2005 by Allison Bellini, a Six Sigma Black Belt who left her secure job as VP of Quality for an appliance manufacturer to develop a unique line of dolls. With over 20 years' experience in consumer products, Allison focused on quality, efficiency, and listening to the Voice of the Customer, when she started All-Belle.

The name of the doll company is a fusion of Allison's first and last name, but it has another special meaning. The motto of All-Belle Dolls is "…because every child is beautiful." All children are beautiful, in the eyes of All-Belle Doll Company. Allison's idea for All-Belle was hatched when she learned that her second daughter would be born with a birth defect. She wanted to prepare her first daughter for her baby sister's homecoming, with a doll that did not conform to the traditional "rules of beauty." Allison searched everywhere for a doll that had features that would help her oldest daughter to become comfortable with facial or bodily features that would be different from her own. She also thought about the day when her youngest child would be ready for a doll and was frustrated that there was nothing in the market to offer her.

Knowing that her personal desire for a doll with unique features was not enough to start a business, Allison spent her vacation time during her last year as a VP of Quality conducting a QFD and using other Design for Six Sigma (DFSS) tools, such as the Kano model. She learned what other parents of special needs children would really like to see in a doll with unique features. Allison interviewed child psychologists, talked with retail buyers, and hired an industrial designer to make drawings of her doll ideas. The printed designs were tested in focus groups and she had samples made, which she tested through ethnographic studies of little girls actually playing with her uniquely designed dolls. The final design was somewhere between realistically molded features, and a fanciful, interpretation, inspired by characters from animated children's movies.

Allison's endearing designs, psychologists' endorsements, and research participant parents' rave reviews got All-Belle off to a sell-out start when she launched her ecommerce website. Over five years, All-Belle gained distribution across the United States and Canada, selling to independent toy stores. The company was manufacturing with two doll factories in China and Allison was pleased that they were both always willing to take her suggestions for quality improvements. She had found many ways to get better pricing over the years, as well, working with the factories on Lean Six Sigma projects. All-Belle Dolls had won several toy innovation awards and Allison had been featured as an innovative entrepreneur by many national media outlets. Despite being started up right before the recession, All-Belle was successful, largely because its founder had a 20-year strategic plan. While success had come quickly, Allison had been careful to take a long-term view for the company's growth, growing a strong team and building her factory relationships.

All-Belle finally landed line review appointments with the top mass merchants and a major big box toy chain in the United States. The buyers all told Allison and her sales team that they were impressed with the All-Belle story and had not seen such innovation in the category in decades. Allison was proud of her young company and her team's achievements in such a short time. All-Belle's quality and unique selling proposition, along with impressive margins for the retailer, won the doll company a number of items on two of the mass retailers' shelves, with her entire line being offered through the retailers' dot-com business.

Allison's excitement over All-Belle dolls being selected for major distribution was quickly replaced by dread when she received the vendor manuals and learned that her factories would have to undergo factory audits (FA) at least 60 days prior to ship date. With purchase orders being issued and ship dates in 120 days for holiday selling season, Allison had little time to prepare her factories.

The retailers' FAs were a little different but, overall, they all included some form of supply chain security (SCS) and factory capability and capacity audit (FCCA), as well as a section on ethical sourcing (ES). Allison knew her factories could pass the first two with flying colors, but she was concerned about the ES audit. It included questions that she had never considered.

Allison started the hard, concentrated work of educating herself quickly on ES requirements. She found that some of these new customers had their own set of ES requirements, while one of the retailers accepted ICTI (The International Council of Toy Industries) audit. The ICTI CARE (caring, awareness, responsible, and ethical) process is the international toy industry's ethical manufacturing program, aimed to promote ethical manufacturing, in the form of fair labor treatment, as well as employee health and safety, in the toy industry supply chain worldwide.

Allison researched the ICTI process and found that her factories would be inspected for compliance in the following areas (ICTI, n.d.):

1. Child labor
2. Prison/forced labor
3. Working hours
4. Wages and compensation
5. Discrimination
6. Working conditions
7. Industrial safety

Allison was confident that her factories were fine with most of these issues. Even though she had always focused on product quality, what the consumer wants, and making sure that All-Belle Dolls offered excellent margins to her retail customers, many of the ES issues were fundamental to Allison. The one issue that she had a level of uncertainty was wages and compensation. Within this area was a question about the factory's compliance with social insurance. Allison was not sure about social insurance requirements. These were questions about the factories' business that she had never considered and was somewhat uncomfortable discussing with their management teams. After a few brief emails and calls, Allison realized that both factories had issues with social insurance compliance.

During the ES audit, there would be multiple stakeholders interviewed, from management to frontline workers. One retailer described the purpose of the ES audit as building a socially and environmentally responsible supply chain by monitoring and strengthening working conditions, community impacts, and environmental practices in the supply chain. This statement brought out concern for another group of stakeholders—those who lived in the local community. Allison had never thought about the communities where her factories operated, or issues like social insurance. She suddenly felt like she did on her first day of Six Sigma Green Belt training, when she realized that there was so much to learn in a short period of time.

Allison had successfully led hundreds of Six Sigma DMAIC projects throughout her career and her doll factories were familiar with this problem solving approach. However, Allison did not think this method would work well to prepare her factories for the ES audits. She decided to instead follow the SOFAIR process.

5.3.1 Stakeholders and Subjects

Allison began by sending the questions for the ES audit to her factory contacts. She asked them to let her know the percentage of employees that are covered by social insurance. The response was troubling; as low as 10% participation in the mandatory social insurance was reported. One factory also explained that the insurance was so expensive that implementing compliance would raise her prices to pay for all employees.

Since Allison now knew that social insurance compliance was going to be an issue, she had questions about what the insurance was for and what were the regulations. Cursory research revealed that there are five social insurances: pension, medical, work-related injury, unemployment, and maternity coverage. The insurance is paid as a bundle, with the pension and medical portions by far the most costly parts of the package. Allison is concerned about the SOFAIR project having such broad scope, but the bottom line is her factories have to comply with all five insurances. To help her better understand how the insurance and healthcare systems are interrelated, Allison, with her top staff members, completes a SIPOS (see Figure 5.6).

She then decides that, to effectively engage with stakeholders, she would need to make a special trip to China. This would be the first time she would travel to her factories without a specific product to develop, quality issue to address, or production to inspect. Before the ES audits would take place, Allison wanted to understand the stakeholders involved with social insurance and understand how the seven subjects of responsibility can be considered.

Allison begins her discovery process to complete a SIPOS. She would normally begin her stakeholder engagement with a factory management meeting. However, she wants to be well-informed for those conversations and she realizes that she knows so little about social insurance that she can hardly complete the SIPOS. Allison has decided to retain the services of a local consultant, who is a former

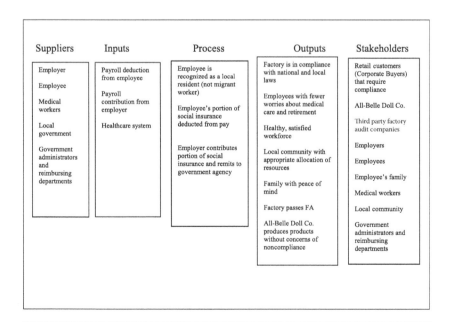

Figure 5.6 Manufacturing example SIPOS—Factory in compliance.

ES factory auditor. In order to gain an unbiased understanding of the problem, before engaging her factories in potentially untenable discussions, Allison accepts the consultant's offer to interview factory workers from other manufacturers, for a first-hand understanding from the employees' point of view.

Allison wants to fully understand the regulations, and she has learned that the regulations are slightly different, depending on the province, country, and even at the city level. Through her discussions with the consultant, the worker interviews and other research, she realizes that social insurance impacts more than the employee themselves, but the local community, as well. Allison also begins to see that there is a real risk to her ability to produce her dolls if workers go on strike. She reads disturbing reports, such as the 2014 strike at one of world's largest shoe factories, a supplier to Nike and Adidas. This strike showed how cost-cutting pressures are squeezing labor-intensive manufacturers and now Chinese workers have become more knowledgeable about their rights (Roberts, 2014).

Allison finds that loopholes in the social insurance system are a problem, and the practice of making less than full payments is common among manufacturers. She is very concerned now, about the potential problems caused by noncompliance with social insurance law. She is concerned about potential disruption to her business if employees strike, but more importantly she is concerned about the local community's ability to respond with any healthcare crisis. Without the correct insurance payments processed for the benefit and investment in the local community, the employee stakeholders making her All-Belle doll may not have the healthcare resources that they need.

Understanding now that the problem of social insurance not being paid is more complex than she could have imagined, Allison then examines the seven SR subjects. She knows that it is important to gain clarity about the full scale of the problem before setting objectives for the project. Most of the workers in All-Belle Doll factories are from provinces far away. The problem is that most workers are relying on the basic insurance that is issued from their home province. But these employees then have no medical insurance in the city where they are working; if they are injured or become ill there, they would have to return home. An alternative is to pay cash for medical care, which is not possible for most low-wage earning workers.

1. Organizational Governance
   a. All-Belle needs to work with factories that are in compliance with Chinese law, as well as the requirements of retail customers. Avoiding full participation in social insurance is a common practice in Chinese manufacturing. All-Belle's factories have their own organizational governance and cultures; we cannot force compliance.
   b. We need to gain more transparency about our factory's payroll and insurance practices. The most effective path will be to gain a higher level of understanding about the organizational challenges to compliance, then influence tactfully.
2. Human Rights
   a. We should ensure that all employees are documented and that employers are not avoiding the declaration of workforce, to avoid paying social insurance.
   b. If employees are not paying into social insurance, do they have difficulty accessing services locally? What happens if they are injured on the job or become ill and they are not registered?
   c. Workers might strike, protest, or in the most tragic cases, take their own lives, to express their dissatisfaction.
3. Labor Practices
   a. We should ensure that employees are participating in the social insurance program, so that their healthcare and retirement needs are provided.
   b. The risk of worker strikes, if the employer does not contribute fully, must be considered.
4. Environment
   a. Labor strikes might disrupt local trash, or other municipal, services, thus creating a local, temporary environmental issue.
5. Fair Operating Practices
   a. We should ensure that our factories are not avoiding paying social insurance as a way to artificially reduce their operating costs and provide more competitive quotes. We must be prepared to source from factories that provide less competitive quotes, but comply with social insurance requirements.
6. Consumer Issues
   a. We were founded on the principle that every child should be able to have a doll that gives them comfort and joy. Consumers trust us to not only do the right thing, but to set the bar higher for what is right. Consumers would likely find our factories' noncompliance to be incongruous with our values.
7. Community Involvement and Development
   a. Nonparticipation in social insurance can cause a city to be allocated fewer resources, from doctors to hospital beds and ambulances, as well as other aspects of infrastructure.
   b. Dissatisfied workers might go on strike and protest, disrupting daily life.

Figure 5.7   Manufacturing example comprehension of the seven subjects.

Though most employees see the benefits of social insurance, they do not benefit from it if their company does not enroll them. Figure 5.7 shows Allison's assessment of the seven subjects.

5.3.2 Objective

Allison now understands the problem better, and the potentially far-reaching impacts. She brings in more team members to formulate the objective of the project. Figure 5.8 shows the All-Belle objective for

All-Belle Dolls, Corp. is committed to working with our China factories to achieve 100% participation in social insurance. Our objective goes beyond the factory passing factory audits with our new retail customers and complying with Chinese law. We believe it is important that the workers in our factories have confidence in their ability to access healthcare and to have peace of mind that they will have a well-funded retirement. We seek to avoid civil unrest or epidemic illness, should the employer not participate fully in the social insurance program. While we understand that there is resistance due to the costs to employers and employees, we are seeking to influence our factories to recognize the importance of participating in the social insurance program.

Figure 5.8   Objective.

this SOFAIR project. As Allison has already noted, All-Belle does not have direct authority over the factories, and the practice of underpaying social insurance is common. How can her team work effectively with factory management to make the required changes?

5.3.3 Function and Focus

Allison has brought her quality team fully into the project by now. As we know, All-Belle has a strong Six Sigma foundation. Her team now understands the complexity of the problem, what different stakeholders want and we know what the All-Belle objective is. Her team remarks that there are many different subjects and more serious risks than they had originally considered. The All-Belle team has a difficult objective; they will need to focus their efforts. Taking the multifaceted problem of social insurance compliance and addressing the different aspects one-by-one is the approach the team agrees upon.

The All-Belle team uses a value function diagram to start to brainstorm opportunities for focusing their problem solving. Figure 5.9 shows our value function diagram, with some failure modes identified. This work was completed with a cross-functional team. Medical and work-related injury insurance rises to the top as presenting the most serious, and shared risks: Permanent disability or illness and no income for an employee who is injured or gets sick. However, Allison's team notes the ancillary failure mode of civil unrest as an outcome of permanent disability with no recompense. Even though the social insurance is bundled, Allison's team decides to tackle the problem of workers not being covered under the medical and job injury programs. This will allow them to focus their efforts and align with community influencers to convince factory management to be compliant with the law.

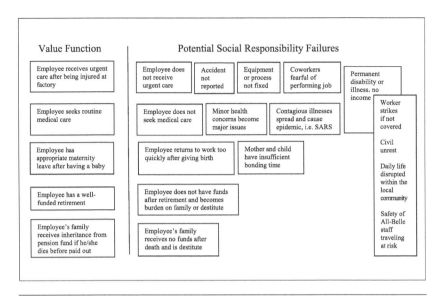

Figure 5.9   Value function diagram.

5.3.4 Analyze

A valuable tool during the analyze phase for getting a system-level perspective of a problem or a process is theory of constraints intermediate objective map, or IO map (Dettmer, 2007). The IO map is a graphical representation of a system's goal, critical success factors (CSF), and the necessary conditions for achieving them. The IO map is intended to fix, in time and space, a firm baseline or standard for what should be happening if a system is to succeed. It represents the desired performance—the destination toward which all system improvement efforts should be directed. The IO map is a tool focused on doing the right things, versus doing things right.

The All-Belle team starts with a system goal of healthy workers. Three CSFs are the employee contributing, the employer contributing, and the local community having sufficient healthcare services (see Figure 5.10). Necessary conditions for the employee contributing include awareness of risks to being an unregistered migrant worker, desire to access local healthcare services, employer support, and sufficient wages. This is where the power of the IO map in identifying system-wide constraints becomes apparent. Any gaps in performance between this desired state and actual conditions can be identified as undesirable effects and expanded to determine root causes.

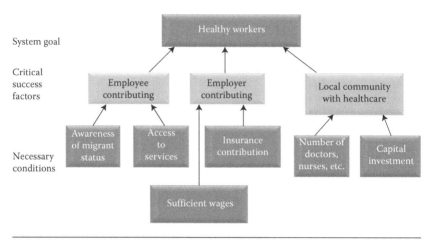

Figure 5.10   Manufacturing example theory of constraints IO map.

5.3.5 Innovate and Improve

We will examine some creativity tools for mistake-proofing against irresponsible behavior through design. However, the problem is largely behavior-based. Behavior is created by both motivation and ability (Grenny and Patterson, 2013). The motivation to not pay insurance is increased profit margin or the award of new business. The ability that leads to the behavior is that nonpayment of social insurance is ubiquitous and there are no negative consequences (yet) for noncompliance. Allison decides to tackle both the motivation and the ability aspects of the behavior.

Although she may not be able to control Chinese politics and culture, she can completely control how All-Belle dolls are manufactured. Recognizing that she can only affect her own business, Allison creates an influence matrix for factory sourcing decisions. Her influence matrix will be focused on All-Belle's sourcing decisions (see Figure 5.11). The use of the influencer (Grenny and Patterson, 2013) model of behavior changes can be an important improvement tool. Grenny and Patterson's model has been used to change complex cultural issues across the globe. Many social responsibility manufacturing issues are deeply embedded in culture. Correcting irresponsible behavior can be difficult. And understanding of the factors that drive behaviors and behavior change are needed.

	Motivation	Ability
Personal	Factory manager fails the factory audit and loses business	Train FA auditors in social insurance compliance expectations
Social	Make known the statistics of worker social media, and the need to avoid factory strikes	All-Belle sourcing agents network with other sourcing agents to discuss and solve social insurance compliance issues
Structural	Introduce social insurance compliance requirements into All-Bell factory quotation process. Notify factories that they will be de-sourced if non-compliance is ever found.	All-Belle employees will conduct social insurance compliance aspects of the FAs. This activity will not be outsourced.

Figure 5.11   Worker health influence matrix.

By focusing on the behavioral systems of motivation and ability, within the narrow focus of All-Belle sourcing decisions she can positively affect the health of some workers in China.

If All-Belle stands firm in its expectations of requiring social insurance program compliance, there is significant motivation in the sourcing system to affect behavior changes. Factory managers stand to lose their jobs if All-Belle pulls its business. All-Belle is motivated to avoid supply chain disruption caused by worker unrest and strikes. Clear communication, both to factories being sources and to All-Belle sourcing employees can enhance motivation. All-Belle employees must be educated in the social insurance compliance issues; networking with other sourcing agents in other companies can help to reinforce and build the ability to find social insurance program compliance issues. And finally, in order to ensure that behaviors changed, All-Belle will need to take ownership of this important auditing function until confidence in compliance is built.

5.3.6 Report and Repeat

All-Belle is now prepared for factory audits. This is a built-in reporting mechanism. However, they decide to expand reporting to include GRI G4 and to start releasing an annual social responsibility report. Included in these will be descriptions of what All-Belle does in the United States to be socially responsible. Repeating the SOFAIR process in the All-Belle factories will be a simple matter of selecting issues that were set aside as outside the scope of the social insurance project. Most importantly, Allison and her team have a more collaborative, transparent relationship with their factories and a greater appreciation for the communities in which they operate.

Through this manufacturing example, we have shown that SOFAIR can be used for very large, complex, and global problems. Many of these problems seem insurmountable. Many social problems have a long history. Although we can't "boil the ocean" in a single project, we can gradually chip away at small, and even entrenched, behaviors. With a continual improvement viewpoint, we can change very large, complex, and global problems incrementally. And over time, the whole world becomes more sustainable.

5.4 SOFAIR Example: Business Process

In this example, we will use SOFAIR to improve the social responsibility performance associated with sourcing information technology code writing. In this example, Balexia is a growing San Francisco-based smart phone application provider. This organization has won the top Global Mobile Award two years in a row. The organization is led by a trio of college roommates, Michael, Brian, and Jennifer. They are in their early 20s. And they never expected to be as financially successful as they are. They each work at least 90 hours a week. Balexia has grown from a start up with no income, to an annual income of $12 million in just the last 18 months.

They are beginning to sense serious work–life balance issues with their employees. Their team is starting to burn out. One of their top code writers almost missed the birth of his son. And they are concerned that another code writer has abandoned his apartment and just moved into the office. Two months ago they reached a peak of 15 code writers on staff; but in the last month they have lost three code writers to various competitors. They not only need to immediately solve the attrition problem and stem the tide of a shrinking staff, they also need to add an additional 10 code writers in the next 6 months to launch the four apps on the drawing board. Balexia decides to use SOFAIR to find a sustainable and socially responsible solution to their problem.

5.4.1 *Stakeholders and Subjects*

Balexia starts their stakeholder engagement with their employees. The owners understand the competitive climate for talent in their market. They want their employees to know how much they are valued and they want to get brutally honest feedback on working conditions. Balexia decides to hire a third-party facilitator to deploy an employee opinion survey. They use this strategy to ensure that anonymity, and thus hopefully honesty, is retained through the feedback process.

The employee survey results in some surprising results. The owners find out that employees are generally happy working up to 90 hours per week. However, what they value is time off when requested; and they want infrequent large periods of time off. Employees don't mind working very long hours when they work; but, they also want to take significant vacations, 4–6 weeks at a time, to enjoy their travel and leisure. Had Michael, Brian, and Jennifer merely acted to reduce weekly hours worked they would have dissatisfied employees and increased, rather than decreased, attrition. They were very happy that they ran the survey.

Because of the focused aspect of the problem (employee satisfaction), the only stakeholders engaged are employees; Balexia wants to focus this project very tightly around the employee stakeholder group. However, they still need to ensure that they have addressed all seven core subjects of social responsibility with their employees. Figure 5.12 shows some of the brainstorming that the owners did with the consultants who facilitated the employee opinion survey. From these survey results, we can see that Balexia employees are socially responsible. They want to be more engaged with their immigrant neighbors, disadvantaged customers, community

1. Organizational Governance
    a. Employees want to be more involved in developing more definitive policies on vacation and time off.
2. Human Rights
    a. Employees want to be able to use time during work hours to volunteer in the community to work with local social workers to improve immigrant conditions.
3. Labor Practices
    a. Employees want longer, sabbatical-type vacations without losing their jobs.
4. Environment
    a. Employees want to be able to use time during work hours to volunteer at the community vegetable garden.
5. Fair Operating Practices
    a. Employees want to ensure that Balexia is operating transparently, accountably, and ethically.
6. Consumer Issues
    a. Employees want to be able to develop some apps or app functions that serve physically challenged customers.
7. Community Involvement and Development
    a. Employees want to be able to use time during work hours to teach junior high and high school students the basics of coding and app development.

Figure 5.12 Employee opinion across the seven subjects.

environment, and local children. They want to make sure that they are participants in organizational governance and that Balexia is operating fairly. And they want long periods of time off. Balexia has used the SOFAIR process to get a good understanding of their employees as stakeholders.

5.4.2 Objective

Michael, Brian, and Jennifer take the stakeholders and subjects information into a session to develop the objective for this SOFAIR project. This will be tough. They have competing objectives. They need to increase their human resources and increase employee satisfaction which means giving employees more time off, which means decreasing human resource capacity. They will need to find a lot more human resource capacity.

For the objective phase of this SOFAIR project, Balexia will need to do some calculations. They need to determine the amount of employee time off, both on sabbatical and on community projects. Then they will need to determine how many more additional employees will be needed for the increased business with the new app development work, but to also compensate for additional employee time off. They determine that they need to double the size of their staff. They need to ramp up to 30 app developers. This will give every employee one month off, without pay (which is fine with the employees according to the survey), per year. The objective of the SOFAIR project becomes: Double the size of human resource capacity. The sabbatical time off becomes a simple policy change; an improvement project is not needed. But, determining how to double the staff size in a very competitive talent market will be difficult; however, maybe less difficult with the attractive change in time off.

5.4.3 Function and Focus

The function and focus phase, like the objective phase, is obvious and easy in this project. The human resource constraint only applies to code writers. Therefore, the only function addressed will be with code writing. The focus of the entire SOFAIR project will be on increasing the capacity for code writing.

This project is a good example of how taking very small incremental improvements facilitates the SOFAIR method. Although, Balexia may have many, many social responsibility performance improvement opportunities, this project has been chartered for a very narrowly focused problem, to make employees happy and thus reduce attrition and increase human resource capacity. We should be able to close this project and start the next one, quickly.

5.4.4 Analyze

Michael, Brian, and Jennifer are excited about the objective phase and the function and focus phase moving so quickly. Maybe a little too confident about how fast this project will be finished. When they start the analyze phase, progress slows. In the analyze phase, they will need to find causes of human resource capacity constraints. Figure 5.13 shows the fishbone diagram of many potential causes. Suddenly, what they thought was an easy project becomes very complex.

Some of these potential root causes are very, very complex. How can a small company like Balexia affect the pipeline of qualified

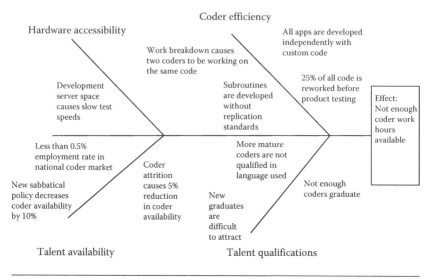

Figure 5.13 Business process example fishbone diagram.

coders? How can Balexia completely change the way their code is written to reduce resource requirements? This may entail a fundamental change to their app platform. In the innovate and improve phase, each of these potential root causes may spark innovative solutions. They decide to carry the causal factors in Figure 5.14 forward to seek solutions.

Problem		Solution
Talent Acquisition	Less than 0.5% employment rate in national coder market	Look outside of national market
	New sabbatical policy decreases coder availability by 10%	Look outside of traditional markets for additional talent
	Coder attrition causes 5% reduction in coder availability	Implement new sabbatical and volunteer policies
Talent Qualifications	More mature coders are not qualified in language used	Deploy retraining process
	New graduates are difficult to attract	Attract new graduates with generous sabbatical and volunteer policies
	Not enough coders graduate	Implement scholarships for code writing skills

Figure 5.14 Causal factors and solutions identified for innovation.

5.4.5 Innovate and Improve

The solutions, identified in Figure 5.14, were developed through some simple brainstorming. After further reflection, Michael, Brian, and Jennifer are disappointed with their solution generation. Many of these solutions have been tried before and they work to a certain extent, but there isn't a lot of confidence that these solutions will be quick to implement or sustainable in the long run. It seems that the need for code writing has far surpassed the supply in the local area for many years. Balexia decides to be more creative in their approach.

They recognize that they are in a field that requires close collaboration and communication, but does not require physical adjacency amongst code writers. They recognize that they have chosen to operate in one of the most difficult local markets for coder resource competition. They realize that their employees are highly motivated to be involved in their community. They realize that work–life balance is difficult, not only for their coders but also for themselves. There are competing forces for and against each aspect of the problem. Figure 5.15 shows a force field analysis of a few aspects of Balexia's problem.

Michael, Brian, and Jennifer do not want a compromise solution. They want to strive for a solution that is so creative that it resolves all of the tension of competing forces. They want a solution that allows for close communication and collaboration and allows for the asynchronous work and significant time off. They want to find a solution that allows for employee involvement in the community and recognizes a culture of long work hours. They want a solution that provides for many qualified coders knowing that there is a limited pipeline of talent.

Factor			Factor
Close communication and collaboration	➡	⬅	Need for asynchronous work and significant time off
Employee desire for community involvement	➡	⬅	Industry culture of excessive work hours
Need for qualified coders	➡	⬅	Limited pipeline of talent

Figure 5.15   Force field analysis.

Balexia realizes that it takes approximately 18 months to train a qualified coder. They do not want to outsource coding to a community in which the company does not operate. They want to foster close employee connection to each other and the community. They decide to relocate the company away from San Francisco. They search for communities that foster community involvement. They look for a community that has a vibrant pool of potential coders, people with a basic understanding of the sciences, easy to train, intelligent, but not necessarily qualified in the necessary coding language. They look for a community that appreciates work–life balance.

After carefully comparing 11 possible cities, they decide to move the company to Chattanooga, Tennessee. They allow the employees that want to stay in the San Francisco area to stay and work remotely. This remote arrangement is only preferred by two employees due to extended family commitments. Since Balexia will keep the current pay rates, most employees jump at the chance to cut their rent in half while keeping pay the same. Balexia works hard to engage employees in community affairs as soon as the move is made. And Balexia immediately launches an initiative with the local community college to develop code writing curriculum that meets their needs. Within 14 months they are resettled, with no attrition, and a growing and learning pool of coding resources.

5.4.6 Report and Repeat

Disruption caused by the move the Chattanooga significantly delays the launch of three of the four new apps to be launched by Balexia. And one of their competitors launches a successful replacement to one of their older apps. It will take Balexia another year to recoup the income losses and regain market share. However, they will do so knowing that their new community and human resource supply is now better and more sustainable. Five years out they are still growing and significantly outpacing their competition in app development speed. They have taken some short-term pain for a long-term gain. Balexia becomes a development supplier for other companies looking for code writing resources. They have turned what was once a constraint into a competitive advantage, all while improving employee satisfaction.

Balexia continues to be a privately owned enterprise; Michael, Brian, and Jennifer have bought out their venture capital investors. Remembering that their employees value transparency, accountability, and ethical behavior, they choose to report on their sustainability performance using GRI G4. And since they are so successful at solving the problem of turning a constraint into a market lever, they are now looking to expand in different industries with similar approaches.

Jennifer has started a spinoff healthcare company. When faced with a dilemma of finding home care for her father, newly diagnosed with dementia, she became aware of the shortage in home healthcare service providers. Jennifer has taken the Balexia solution and is now applying it to the healthcare industry. She recruits healthcare workers from very rural areas, provides them with training, certification, and transportation, and very flexible hours so that they can still take care of their families. Some of her employees choose only work 10 hours per week. Employees are satisfied and she has found a solution to a very tight talent pool. A sustainable, creative solution in the information technology industry has successfully translated to the healthcare industry.

5.5 SOFAIR Example: Personal

This example is about using the SOFAIR method to improve our social responsibility as an individual. Not only should we set standards for social responsibility for the organizations we work for or buy products from; but, we should also live authentically with the principles of social responsibility. There's an added benefit. If we practice CISR in our homes, we will be more adept at using the methodology in our workplace. The following example considers a dilemma that many face. Let's say that we have elderly parents in need of care. Our father is having issues with dementia; our mother is having mobility issues. Between them they can get along independently, with effort. If either has a serious issue, then care for the other parent is immediately needed. Our goal is to keep our parents living independently in their home as long as is practical and to have ready resources to help if there is an accident or healthcare crises. Let's see how SOFAIR can help us to find the most sustainable solutions for all the stakeholders involved in this problem.

5.5.1 Stakeholders and Subjects

The very first step of SOFAIR is to identify and engage the stakeholders of the process. So we need to determine what the process is and who the stakeholders are. We can use the SIPOS tool for this (see Figure 5.16). The stakeholders we will want to engage with are members of the immediate family, the extended family, friends, neighbors, and the community. Figure 5.17 lists some questions we might ask these stakeholders to get the dialogue started. Through this dialogue, we are trying to ensure that the interests of all stakeholders are known.

Through this process, we find that the immediate family and extended family are not easily able to provide full-time care for our parents. In a crisis situation, many family members have indicated that they could step in to help for a few days, but not much longer. We also find through the stakeholder dialogue that many friends and family have had to resolve the same type of situation. We will want to learn from their experiences. And we find out from the dialogue

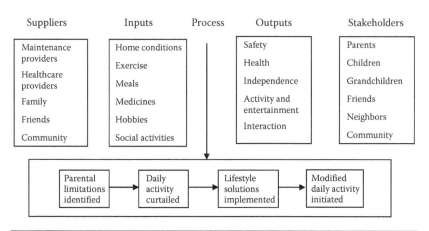

Figure 5.16   SIPOS.

1. Do you have elderly parents?
2. Do they live independently? If not, how do they live?
3. How did you make decisions concerning their welfare?
4. What resources did you use to help?
5. What concerns do you have for my parents?
6. Do you think they should live independently? If so, why?
7. What advice would you have for me?

Figure 5.17   Stakeholder dialogue questionnaire.

1. Organizational Governance
    a. We should make this decision with the involvement of the whole family.
2. Human Rights
    a. We should ensure that all healthcare and maintenance workers have legal work status.
3. Labor Practices
    a. We should ensure that all healthcare and maintenance workers have appropriate working conditions, compensation, and work hours.
4. Environment
    a. We should ensure that all excess food and medicines are appropriately disposed.
    b. We should ensure that the house is maintained for energy efficiency.
    c. We should ensure that any lawn care services are environmentally safe.
5. Fair Operating Practices
    a. We should ensure that any care facilities operate fairly.
6. Consumer Issues
    a. We should ensure that our parents are protected against financial scams and identity theft.
7. Community Involvement and Development
    a. We should ensure that we use community care services when available.
    b. We should ensure that our parents stay active in the community.

Figure 5.18 Business process example comprehension of the seven subjects.

with the community that a nearby senior center provides advice and counseling on making decisions about living independently or with full-time nursing care. This was fruitful dialogue.

Now we need to turn our attention to the subjects. Figure 5.18 demonstrates our notes from our consideration of each of the seven subjects. This is just a quick bit of brainstorming on our top level concerns across all seven subjects. Even for an issue that only involves immediate family and friends; we can see that elements of all seven subjects are potential concerns. We want to develop solutions to our parents' care that are as responsible as possible across all subjects.

In this phase of the SOFAIR method, we are assembling and absorbing information. We want to gather information about our stakeholders and we want to synthesize stakeholder concerns with all seven subjects of social responsibility. At this point, we aren't even sure if we'll use all of the information we've gathered. We are like sponges, using divergent thinking to conduct inquiry on our stakeholders and subjects.

5.5.2 Objective

Now that we've been well advised, in the objective phase we want to begin to focus more specifically on the issue at hand. We create an objective for this project. What outcome is ideal? What are our goals? Here is the objective that the team created:

We want our parents to be safe, healthy, active, and happy. We want them to be together. We don't want them to live in a way that prevents their children or grandchildren from also living a safe, healthy, active, and happy life. We want to ensure that they have a living arrangement that will be comfortable and enjoyable.

This is a simple objective statement for a very important and complex problem. It encompasses our vision of the ideal outcome for our SOFAIR project.

5.5.3 Function and Focus

Now we know what everyone involved in the decision needs and wants. We understand that there are many different subjects to consider with our decisions. And we have a goal, a vision, and an objective for our efforts to help our parents. In the SIPOS (Figure 5.16), we have a simple process flow diagram. Parental limitations are identified and then lifestyle activities are curtailed. Then solutions to the lifestyle issues are found and modified activities can be initiated. This process cycles over and over again. We should anticipate that as our parents continue to age that new limitations will arise and this process will need to cycle again. Therefore the process step, or function, upon which we should focus, should be that of finding lifestyle solutions. We should build decision-making systems that are easy to engage repeatedly as new lifestyle limitations arise. This becomes our focus. Our focuses for this project is how to rapidly implement new lifestyle solutions as new limitations arise.

5.5.4 Analyze

In the analyze phase of this project, we need to understand the inputs and outputs of finding new lifestyle solutions. What information and activity is needed to find new lifestyle solutions? And, what information and activity is needed to implement new lifestyle solutions? In this phase, we are thinking about all of the inputs that are needed to find a solution, and all the inputs that are needed to implement a solution. For this project, the process is not complex. The analyze phase will be very short.

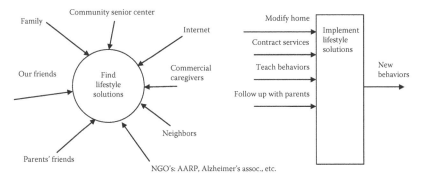

Figure 5.19 Cause and effect of finding and implementing lifestyle solutions.

We find that there are many contributing factors that can be considered inputs to a lifestyle solution system. Family, friends, parents of friends, neighbors, the Internet, community senior center, NGOs, and commercial care givers can all be inputs to a solution. We also identify a system of inputs to instill new behaviors: modify the home, contract services, teach new behaviors, and follow up with parents. In the analyze phase, we take this new understanding of causal factors on solutions and behaviors and create a procedure for finding lifestyle solutions (see Figure 5.19). In this example, we have a nine-step process to cycle through every time there is a new solution needed. This presents a dynamic process of finding a solution every time a new lifestyle limitation arises for our parents.

5.5.5 Innovate and Improve

Our goal is not to find solutions. We are looking for systems to find solutions. Figure 5.20 shows a process flow diagram which will be our new solutions finding procedures. Anytime a new limitation arises for

Step	Procedure
1.	Identify the limiting behavior.
2.	Implement immediate containment for the risks associated with the behavior.
3.	Seek solutions that may already exist for this behavior using inputs from all sources.
4.	If no ready solution exists, use innovation tools to create a new solution.
5.	Test the solutions for effectiveness.
6.	Test the solution against SR risks of all seven subjects.
7.	Test the solution for satisfaction of parents.
8.	Implement the solution.
9.	Follow up on effectiveness and satisfaction with the solution.

Figure 5.20 Procedure for lifestyle solutions.

Step	Procedure	Example
1.	Identify the limiting behavior.	Father walks out of the house and forgets how to return home.
2.	Implement immediate containment for the risks associated with the behavior.	Internally keyed padlocks are added to the front and back doors. Mother is given the key on a necklace. This is a very undesirable solution. In the event of an emergency, egress from the house can be difficult. It is only implemented as an emergency, short term solution.
3.	Seek solutions that may already exist for this behavior using inputs from all sources.	Input is sought from friends, family, and neighbors. An internet search results in finding GPS tracking devices that can be permanently attached to father. Some even have alarms to alert Mother, and others, when Father has left the house.
4.	If no ready solution exists, use innovation tools to create a new solution.	A ready solution exists.
5.	Test the solutions for effectiveness.	A device is purchased, attached, and activated as an ankle bracelet.
6.	Test the solution against SR risks of all seven subjects.	The device passes our concerns for all seven subjects.
7.	Test the solution for satisfaction of parents.	Mother is very happy with the solution. The ankle bracelet does not cause any discomfort or chafing with Father.
8.	Implement the solution.	The solution is implemented. Additionally, a niece who lives a few blocks from our parents agrees to have her cellphone alarm if Father leaves the house.
9.	Follow up on effectiveness and satisfaction with the solution.	Six months after implementation, the device is still working properly. Batteries are changed and a new band installed.

Figure 5.21 Father's example procedure for lifestyle solutions.

our mother or our father, we can initiate this procedure to find a new solution. Figures 5.21 and 5.22 show two examples of exercising this procedure. The social responsibility key to this improvement is ensuring that all inputs are involved in finding the solution and ensuring that all solutions are tested against the seven subjects. And effective, but unsustainable solution is not desirable.

5.5.6 Report and Repeat

As a personal example, there is no public reporting to be made with this example. However, a note of status and gratitude is sent to the immediate family and they are apprised of our parents' status. We will continue to use the procedure as their condition changes, sometimes day-to-day. We can also modify this procedure, slightly, for many other family situations, such as making decisions about taking a new job, new responsibilities for young children, or other family issues.

Step	Procedure	Example
1.	Identify the limiting behavior.	Mother's physical mobility issue has resulted in muscle atrophy that is causing balance issues. She has slipped in the bath. No significant injuries, but improvement in her balance needs to improve.
2.	Implement immediate containment for the risks associated with the behavior.	A home care nurse is contracted three time a week to help Mother bath. She is instructed not to take a bath without her nurse.
3.	Seek solutions that may already exist for this behavior using inputs from all sources.	Input is sought from friends, family, and neighbors. The community senior center has a strength training program targeted for balance improvement. It takes participants with the same mobility limitations as Mother.
4.	If no ready solution exists, use innovation tools to create a new solution.	A ready solution exists.
5.	Test the solutions for effectiveness.	Mother is enrolled in the strength training classes. However, the nurse is also retained until the senior center physical trainer ensures that Mother has built the strength necessary to improve her balance.
6.	Test the solution against SR risks of all seven subjects.	There is a potential human rights concern if our nurse does not have legal employment status. We confirm with the healthcare provider that they use a verification system for employment status.
7.	Test the solution for satisfaction of parents.	After 5 week of training, Mother's strength has improved and the nurse is released. Mother's strength also allows her to move better through the house and garden. She is very satisfied with the results.
8.	Implement the solution.	Mother will continue the strength training indefinitely. This has also improved her social connectedness. She has new friends from her fitness class. Father goes with Mother to the senior center during her class, this too has improved his social connectedness.
9.	Follow up on effectiveness and satisfaction with the solution.	Six months after implementation, the strength training is doing well. However, a new knee pain issue has arisen and the orthopedic doctor will need to assess Mother's ability to continue.

Figure 5.22 Mother's example procedure for lifestyle solutions.

This has been an interesting example. Previously, we saw business-oriented examples in healthcare, manufacturing, and business processes. In this example, we can see that SOFAIR can also be applied to personal issues. The people that surround us in our personal lives are stakeholders. They are impacted by our decisions and actions. Simply following the SOFAIR method, in consideration of our family, friends, and neighbors as stakeholders, can create sustainable solutions and opportunities related to these relationships.

6

TAKING ACTION

We've now shown you several real-world examples of how to use SOFAIR. All of these examples represent small projects, of refined focus, that take an organization (including a family as an organization) one step closer to a CISR®* culture. Remember, that's the goal. Not the accomplishment of projects, but the change of a culture. It is only when everyone in the organization thinks and behaves with a 1500 year view point that a socially responsible culture has taken place. In addition to the accomplishment of SOFAIR projects, within a CISR program, there are many, many more actions that people in the organization can take to promote a culture change.

In the following paragraphs, we detail things that the organizational leader, project leader, communicator, and team member can do to help transform the organizational culture to a CISR culture. These are all easy actions taken by the individuals in their roles. Organizational leaders have directorial opportunities to take action to change the culture. SOFAIR project leaders have actions, in addition to project leadership, that can promote culture change in the organization. Communicators, those people fulfilling roles in the organization focused on internal and external communication, can be very influential in garnering culture change. And finally, we discuss how the individual team member (this includes ultimately everyone else in the organization) has actions that he or she can take, immediately, to advance a CISR culture.

6.1 Ten Things an Organizational Leader Can Do Today as Social Responsibility Action

It is important that CISR is actionable. Action can take place at many levels of an organization; and action can take place now. We want to

---

* CISR® (sounds like scissor) is a registered trademark and can be used with permission for non-commercial use. Contact SherpaBCorp.com for permission.

wrap up the book with the most important step in any CISR initiative: action. Hopefully, we've helped increase your knowledge on a set of tools and techniques, and served to ignite your passion for sustainability; now it's time for action. Let's explore what the organizational leader, the project leaders, the communicator, and the team member can do to move to action now. The following lists and direction can be readily adopted before, during, and after a full CISR program is deployed.

Behavior starts with awareness. There are some basic behaviors of which organizational leaders should be aware when it comes to social responsibility. Leaders should know ISO 26000 and the seven principles and the seven subjects. Leaders should be aware of who their stakeholders are and the need for stakeholder dialogue. As for CISR and SOFAIR, leaders should recognize that continual improvement is a journey; perfection will never be achieved because the stakeholders' expectations will always be increasing. They should be aware of the rigor of a methodology. And they should be aware of the need to just start somewhere. But if awareness doesn't lead to action it's of no use. So the leader needs to initiate and motivate activity. Here are 10 actions that the organizational leader can do now to advance culture change:

1. Set the tone with the seven principles
2. Set strategy with 1500 year thinking
3. Embed SR as a part of business strategy
4. Measure progress
5. Allocate resources
6. Set SR training and development expectations
7. Reward and recognize SR performance improvement
8. Benchmark and collaborate
9. Persevere
10. Walk the talk

6.1.1 Set the Tone with the Seven Principles

The seven social responsibility principles are accountability, transparency, ethical behavior, respect for stakeholder interests, respect for the rule of law, respect for international norms of behavior, and respect for human rights. Begin to live these principles. You aren't

perfect; be accountable to your mistakes. Make your decision methods transparent. Always behave in consideration of your impact
on others. Respect your stakeholders. Follow the law. Understand
United Nations compacts and norms. Recognize and respect
inalienable human rights. Make your life and behavior socially
responsible.

For example, let's take an information technology (IT) manager
in a mid-sized corporation. This role is what we would consider an
organizational leader. The IT manager may not have a direct role in
the CISR program of her organization. However, in her function,
she can set the tone that the seven social responsibility principles are
important. And she can lead her IT organization in a socially responsible way. She can develop hardware and software sourcing decisions
that are transparent and accountable to the whole organization. She
can ensure that audit mechanisms are deployed in the IT function.
She can ensure that sourcing decisions respect the rule of law, respect
international norms of behavior, and respect human rights. She can
ensure that these principles are abided by with respect to her off-shore
code writing supplier. And she can ensure that members of her organization recognize all of their stakeholders.

6.1.2 Set Strategy with 1500 Year Thinking

Lead your organization with future generations in mind. Answer the
question: What will this organization's objective be in 1500 years?
Then work to set the foundation for success in that timeframe. Will
you have customers in 1500 years? Will you have suppliers, raw materials, and capable employees?

This action requires a cohesive organizational strategy that considers a very long-term outlook. Go back to our Kongo Gumi example in
Chapter 1. Kongo Gumi had to be concerned that there was a source
of wood for building materials for 1500 years. Sustainable forestry
was required. Kongo Gumi had to ensure that it had skilled carpen-
ters for 1500 years. Employee and community development needed to
be sustainable. Kongo Gumi had to have Buddhists for 1500 years. As
temple builders, the worshippers were their customers. Community
involvement was required on the organization's part. Kongo Gumi
had to have leaders of the company for 1500 years. In this example,

Kongo Gumi innovated to allow the husbands of daughters to become family members to ensure male lineage for organizational leadership. Their organizational structure and governance is sustainable. Your organization should be taking action in consideration of 1500 years of sustainable stakeholder engagement.

6.1.3 Embed SR as a Part of Business Strategy

Do not consider your social responsibility efforts as an addition to profitability or business success; see them as mutually inclusive. Operate responsibly *in order to be successful*, not in addition to being successful. Do not seek initiatives for spin or marketing benefit. Make your initiatives integral to your business success.

Let's consider, for example, Mark, the CEO of a mid-sized food company that specializes in nutritionally balanced snack food and drinks for geriatric customers. The company's business strategy is focused on sales growth and product safety. Both of these business strategies can be used to promote social responsibility. Think people, planet, profit; all three achieved simultaneously. The company's sales growth strategy could be targeted at the most vulnerable stakeholders. The company could engage employee stakeholders in volunteering their time with geriatric community organizations; employees become the face of the company in the community improving brand recognition and thus increasing sales. The company could initiate innovative design projects to reduce packaging waste thereby improving environmental impact and reducing cost, leading to lower prices and increased sales growth. These are examples of how sales growth initiatives can promote social responsibility. The company can initiate projects to ensure suppliers use organic raw materials, therefore improving product safety by eliminating health-fragile customer ingestion of pesticide and fertilizer residue. The company could work to source raw materials as close to the manufacturing sites as possible. This would reduce the risk of raw material spoilage, reduce freight costs, and ensure fair labor practices and human rights considerations by suppliers. These are examples of how product safety initiatives can promote social responsibility. Achieve successful business strategy through the use of continual improvement for social responsibility.

6.1.4 Measure Progress

Adopt sustainability reporting practices. Know your metrics, report your metrics. Make sure that your social responsibility improvement efforts yield measurable results. You cannot prove you improved without measuring performance. Usually organizations start measuring social responsibility performance by adopting a standard set of metrics such as GRI G4 (The Global Reporting Initiative, n.d.). Some standard metrics will be beneficial to the organization, some will not. Some standard metrics will be reactive (meaning that they measure a result after the fact) and some will be proactive (meaning that they predict an outcome). Using a predetermined framework is an easy way to start. Later, your organization should strive to choose social responsibility performance metrics that are more and more predictive and proactive, and more and more customized to your business strategy and stakeholder needs.

Following the last example, the geriatric nutritional food company may choose to measure total water withdrawal by source (a standard disclosure for GRI G4) as an environmental protection-oriented social responsibility metric. Based on the results of measuring this performance, the company may then initiate a SOFAIR project that leads them to innovate their nutritional drinks to now be shipped as a dry powder for the customer to add water. This innovation decreases shipping costs, reduces water waste, and makes the product easier to carry home from the market for the customer. Measuring water withdrawal by source may now be a metric that does not necessarily initiate continual improvement for social responsibility, and a new metric needs to be implemented to promote new improvement.

6.1.5 Allocate Resources

The CISR program will need project leaders, trainers, travel, and team member time. In any continual improvement initiative there is surge capacity that is needed to affect change. After the change, fewer resources will be needed; but during the initiative more resources will be needed. The organizational leader should take action to provide for these resources. Will the CISR resources be engaged as third-party consultants? How many project leaders are needed on an ongoing basis? What is the capacity of organizational team members to absorb the project work load? These resource questions will need to be answered and the program will need to be supported by the allocation of resources.

6.1.6 Set Social Responsibility Training and Development Expectations

Most organizations have professional development expectations for employees. Include CISR-related development expectations for your employees. For employees who will be team members on SOFAIR projects, perhaps the training will happen on the project. Project leaders may require formal classroom training on the SOFAIR methodology. And organizational leaders may require some training on social responsibility and how to be a good SOFAIR project champion. Organizational leaders will need to develop the answer to these questions: How fast, will members of the organization be trained in the SOFAIR methodology? Who will be trained? What training facilities will be needed? Will internal or external trainers be engaged? How will training be sustained in an ongoing way?

Additionally, social responsibility professional development opportunities can be adopted. Project leaders can be expected to benchmark social responsibility practices with other organizations. Conferences and seminars can be attended. Expert level training and certification can be achieved. The organizational leader can move to action by setting these professional development expectations.

6.1.7 Reward and Recognize SR Performance Improvement

With the deployment of a CISR program, and with many SOFAIR projects being completed in different areas of the organization, results will start to be achieved. These results should be widely publicized and recognized. Recognition events held to publicize the great work of SOFAIR project teams should be planned. Recognized performance should be connected back to business strategy and performance metrics. Consider the SOFAIR project team members as employee stakeholders; make recognition customized to the desires of the team. Social responsibility performance improvement should be broadly publicized and recognized.

Employees are attracted to organizations that promote social responsibility (Santos, 2013). Singing the praises and promoting awareness of the organization's social responsibility success may improve employee attitude and retention. It is our opinion that rewards associated with social responsibility performance improvement be socially responsible. For example, reward social responsibility performance improvement by allowing team members to spend extra work time contributing to a favorite charity. Reward social responsibility performance by allocating resources to community outreach events. Include families in the reward and recognition; they are stakeholders, too.

6.1.8 Benchmark and Collaborate

As an organizational leader you should know what your competition is doing. You should know the latest trends and techniques in your industry. You should have a vibrant network of thought leaders and experienced mentors. This will require that you take action to get out of the "shop" on occasion and look around. Benchmark and collaborate with other organizations.

For example, in our example company that provides nutritional food for geriatric clients, they may seek to collaborate with retirement and nursing homes. They may initiate joint projects with these customer stakeholders to improve the delivery of their product to the geriatric clients. They may benchmark food products in hospitals and other institutional settings. They might benchmark the social responsibility innovations of baby food producers. They might collaborate with geriatric fitness gear producers or physical therapists. Get outside of the organization to see what else is going on.

6.1.9 Persevere

The average organizational culture change takes many years (Katzenbach et al., July–August 2012). Creating a socially responsible organization is a marathon, not a sprint. This effort will take years or decades, not months. One approach to such a challenge is to break it into smaller steps. Set up a 3-year plan, a 5-year plan, and a 10-year plan. These strategic plans should answer questions such as: How will you know, over time, that the culture is changing? How will you know that the organization is on pace? How will the breadth and depth of the culture change be visible over time? Who will succeed you when you are no longer the organizational leader and will your successor continue your work?

The ultimate goal is for the social responsibility *program* to disappear. You will know the culture has changed when all seven principles and all seven subjects permeate everything that the organization does. Being socially irresponsible is unacceptable. And you will sustain the efforts during leadership succession if the base of the organization has already achieved a culture change. Therefore, deep and broad changes are necessary. What you can do now, as an organizational leader, is to involve as many people as possible, at the base of the organization, in SOFAIR projects. And keep doing projects. The minute a project leader finishes one project, he or she starts another. If everyone in the organization becomes involved and engaged in incremental improvement, through SOFAIR projects, not once, but continually, culture change will happen.

6.1.10 Walk the Talk

Are you leading authentically? Do you say one thing and do another? Do you exhort energy conservation, yet keep your office lights on when you aren't in the room? Do you promote human rights considerations, yet have housekeepers or gardeners at home with sketchy immigration status? Do you expect your organization to engage with community stakeholders, yet isolate your family from the same community? The biggest action that a leader can take in any situation is to live his or her beliefs, to be authentic to the core, in what he or she publicly promotes. The most important action that an organizational leader can take right now is to become socially responsible.

6.2 Ten Things a Project Leader Can Do Today as Social Responsibility Action

By using the SOFAIR method, performance is improved incrementally, project-by-project. Therefore, the project leader plays a very important role. In some organizations, these project leaders may reside in a continual improvement organization. They may work in a process engineering group, a quality assurance or quality management team, or, if the organization is large enough, a sustainability or social responsibility department. In general, when we say project leader, we mean anyone, anywhere, in the organization that is leading social responsibility performance improvement, with or without the SOFAIR methodology.

1. Bring stakeholder awareness to the project
2. Use what works for you
3. Motivate team members
4. Engage cross-functionally
5. Engage with partners inside and outside your organization
6. Share with other project leaders
7. Communicate project results
8. Continually improve your skills
9. Be aware of risks
10. Manage your projects

6.2.1 Bring Stakeholder Awareness to the Project

An extremely important aspect of social responsibility is stakeholder consideration and stakeholder dialogue. We tend to miss identifying risk when we do not fully understand the potential losses that could be important to certain stakeholders. Thorough stakeholder dialogue allows us to have better information to ascertain risk. And it is very easy in the daily hustle and bustle to forget about some of our stakeholders. The project manager can ensure that awareness of concerns from all stakeholders becomes a concern for the project.

For example, let's say that we are working on a project to improve the waste in our product packaging. We want to reduce our carbon footprint and shrink our impact on contributing to the solid waste stream. An effective SOFAIR project leader will conduct thorough stakeholder dialogue as a part of the stakeholders phase. One stakeholder group is employees. Our example project manager gets feedback on how unreliable the current plastic welding machine is in the current packaging process. Improvements to the packaging, especially those that eliminate plastic and plastic welding, will not only help the environment; it can improve the productivity, motivation, and job satisfaction for our employees. Assuring awareness of all stakeholders impacted by the process, and the process change, is valuable action taken by the project manager.

6.2.2 Use What Works for You

Many SOFAIR project managers have project management expertise in areas other than sustainability or social responsibility. Perhaps they have experience in construction, engineering, or other process improvement projects. Experience gained from these other functions and professions helps the SOFAIR project. If we take the perspective that our primary goal is to prevent losses from socially irresponsible behavior, then CISR is really just loss prevention through risk abatement. Many professional disciplines deal with risk analysis and loss prevention. Auditing, engineering, quality assurance, accounting, and many other professions have tools and techniques for risk analysis and loss prevention. For SOFAIR project leaders coming from these types of professions, use what you know; use what works for you.

For example, one type of risk assessment tool from the field of accounting is promoted by the American Institute of Certified Public Accountants (Landes, 2012). This is a subject assessment checklist type tool used to identify significant business risks that may result from the important mistakes in a client's financial statements. Someone from an accounting background might find that it is easy to revise this checklist to identify social responsibility risk. Go for it. Use what works for you. Innovation in CISR methods can only move us all forward.

6.2.3 Motivate Team Members

Project leaders should be interested in motivating their team members to champion social responsibility at work, on the project, and at home, throughout their lives. Breaking down large-scale, difficult, and chronic social responsibility challenges into discrete, incremental projects can help everyone understand that valuable improvement can be made by anyone at any time. And recognizing this agency can motivate action.

For example, let's say that our project leader is operating in a location where governmental corruption is a social problem. The project leader is leading a small-scale project to improve the social responsibility of transporting material between two sites, both belonging to his employer. The truck drivers (who are employee stakeholders) state that they are regularly getting stopped by the same police officer who requests a small bribe to avoid inspection of the truck load. Other issues associated with this project include the environmental impact of the transportation, safety concerns with the truck traffic, community safety issues from the truck traffic, as well as fair labor issues with the treatment of the truck drivers. As a part of the project, the project leader considers and includes the stakeholder input on the governmental corruption. This helps the team members begin to realize that, maybe only in a small way, they have an opportunity to chip away at this chronic problem. The team comes up with a solution to combine the movement of material from site to site with the trucks that are transporting finished products. These trucks are international, regulated, and certified trucks. The team decides to stop paying the bribe and allowing the inspections to take place, with the improved controls on the trucks and their respective safety. In a very small way they have improved the chronic problem of corruption in their community and improved the employees' confidence in personal agency for change.

6.2.4 Engage Cross-Functionally

One of the benefits of team-oriented problem solving methods, such as SOFAIR, is the inclusion of people from many different organizational functions on the team. Partnering people from customer service, operations, accounting, human resources, and engineering provides for diverse perspectives and creative approaches. Engaging cross-functionally improves creativity and innovation. The project leader can build cross-functional collaboration to achieve improvements in social responsibility performance.

6.2.5 Engage with Partners Inside and Outside Your Organization

Initiating stakeholder dialogue is a form of engaging partners inside and outside your organization. Stakeholders, by definition, are any individual or group that has an interest in any decision or activity of an organization. If they are interested, they might make a great partner. However, there are also partners that may not be stakeholders. For example, NGOs might be interested in facilitating collaboration between groups without being a stakeholder to either group. These might become partners from which social responsibility improvements can arise.

One example of a helpful partner might be a professional society. Previously, we used an example of the association of certified public accountants. Although their risk assessment tool was not created with a SOFAIR project in mind, and the association membership is not focused on social responsibility improvements, but, engaging with the organization to understand their risk assessment tools can help the SOFAIR projects. Project leaders who cast their nets widely to engage with partners inside and outside their organizations can gather great ideas to apply to their projects.

6.2.6 Share with Other Project Leaders

Although the methodology and toolkit are robust, each SOFAIR project is slightly different from the last. And each project will have a new set of feedback from stakeholder dialogue. There is a lot to be learned by comparing project work. Sharing with other project leaders is a valuable action for any project leader to take. Comparing stakeholder feedback, risk assessment results, improvement ideas, measurement and reporting metrics, and innovation and process change actions can enhance the project leader's understanding of the organization, stakeholders, and best practices. The project leader should approach CISR as an ever-learning opportunity. Sharing with other project leaders is an action that should be initiated from the start.

6.2.7 Communicate Project Results

Projects can be expensive endeavors. There are the cost of the team members' time, the cost of training project leaders, and the costs of the process improvement. The benefits will outweigh the costs, but the costs should be a consideration. For this reason we never want to learn the same lesson twice and we never want to solve the same problem twice. Communicating project results ensures that process improvement can be replicated without the expense of another full project.

For example, let's go back to the packaging improvement team. The packaging design improvements that eliminated plastic and plastic welding should be evaluated as applicable improvements to other projects, at other sites, and in other product groups. There should be no need to conduct another packaging improvement project to solve the same problem all over again. The improvements from the original project can be immediately replicated for more rapid social responsibility performance improvement, and at a lower cost, since the cost of the SOFAIR project is no longer a burden.

6.2.8 Continually Improve Your Skills

Each project will improve the project leader's skill set, tool set, leadership skills, organizational influence, and confidence. Formal training, in addition to these on-the-job skill enhancements should be continually considered. Training in project management skills, team leadership skills, and social responsibility issues and solutions are some good options. A development plan for SOFAIR project leaders should be a standard practice. Getting outside of the organization and going to a conference may be another way for the project leader to sharpen her saw. Special projects within the organization can improve project management skills. Speaking to different stakeholder groups, such as investors or employees can sharpen leadership skills. Participating in community outreach efforts or volunteering for leadership positions with professional societies or other nongovernmental organizations can expose the project leader to broad societal issues and their solutions. Continual improvement of the project leader's skills is a CISR objective.

6.2.9 Be Aware of Threats

The CISR approach to social responsibility is one of taking preventive action. We don't want to wait until irresponsible actions happen before we effect system change. Therefore, the project leaders should always be on the lookout for new organizational threats. The business environment is always changing. Stakeholder expectations are always changing. There will always be new risks. Being vigilant about threats awareness is an important action with which the project leader can be involved.

Going back to the packaging improvement example, the project leader that led that project is now an organizational expert at packaging design. The level of involvement required during the projects results in the situation that project leaders never really completely separate themselves from the project topic. Let's say that the SOFAIR project team removed all the plastic from product packaging and replaced it with cardboard. Two years after the completion of the packaging project, the same project leader becomes aware, through his professional society engagement, that there is a wide off-shore issue with contaminated cardboard. This is a new risk to the organization, and one in which he has the ability to contact those in the organization who can immediately react to this situation. Continual improvement on the product packaging is needed to ensure that off-shore supplied cardboard is tested for the toxins. This is an example of how heightened risk awareness by project leaders can continue to promote a CISR culture.

6.2.10 *Manage Your Projects*

SOFAIR projects are an investment. There is an investment in time, money, organizational attention, and stakeholder engagements. Exercising the full tool kit can be difficult and complex. Solving social responsibility challenges takes effort. This effort is not to be taken lightly. The project leader should be a good project manager. She should use her project management tools for effective project management. The SOFAIR methodology, analytic tools, work breakdown, resource balancing, critical path analysis, project planning, and verification of action completion are actions required by the project leader for effective project management.

6.3 Ten Things a Communicator Can Do Today as SR Action

The role of communicator can be filled by many different positions in the organization. This might be someone serving in the function of corporate affairs, investor relations, or public relations. These are people, primarily, employed by the organization for external communications. But, they might also initiate internal communications as well. The training in communication, investigation, social media, and public relations that these professionals have received makes them great candidates to lead stakeholder engagement activities. The actions listed below serve as ideas on how to use the communications skill set of these important team members to improve social responsibility performance.

1. Raise awareness
2. Report—publicly
3. Facilitate stakeholder dialogue
4. Stop the spin
5. Look for active and passive stakeholder information
6. Connect people and initiatives within the organization
7. Balance business needs with reporting
8. Connect any philanthropy to business strategy
9. Be a participant in projects
10. Communicate praises of social responsibility improvers in the field

6.3.1 Raise Awareness

Larger organizations might have communications mediums such as internal and external websites, internal newsletters, press releases, and social media platforms. This can also include not just broadcasting one-way communication, but also receiving information such as internet feedback analysis, focus groups, marketing or industry communications analysis, and customer feedback. In some organizations, these same team members are responsible for sharing the organization's social responsibility and sustainability efforts, up to, and including, formal annual sustainability reports.

All of these internal and external communications, formal and informal, can be opportunities to raise stakeholder awareness about social responsibility. The communicator in the organization has a ready-made platform to raise social responsibility awareness amongst all stakeholders reached. This should not only include information on what the organization is doing to improve social responsibility performance, but it can also include information, tips, and techniques that other stakeholders can take to improve their social responsibility performance. For example, employee newsletters could include home energy conservation tips, thus positively affecting the environmental impact from every employee. Social media feeds could inform customers on subtle, but important, social responsibility benefits to choosing the organization's product as compared to industry competitors. The company's website could offer details on how social responsibility principles are adopted in the organization and, thus promote those same principles in potential suppliers. The communicator can use the existing communication platforms to raise the general awareness of social responsibility with all stakeholders.

6.3.2 Report: Publicly

Many organizations that are dedicated to social responsibility performance improvement use an internationally recognized standard as a starting point for public reporting on the status of sustainability efforts and metrics. For public corporations this reporting might be integrated with the summary annual report for the whole business, or it might be published as a standalone sustainability report. The Global Reporting Initiative (n.d.) and the Investor Responsibility Research Institute (n.d.) are two examples of internationally recognized public reporting frameworks and standards. Regardless of the type of reporting standard used, frequency, or content, reporting publicly demonstrates the organization's prioritization of social responsibility.

Reporting publicly supports the social responsibility principles of transparency and accountability to very large stakeholder groups. Public reports also hold important information for future potential stakeholders, such as future employees, customers, and suppliers. Public reporting can give the local community information on the organization's social responsibility efforts and feedback that their involvement in stakeholder dialogue is important. Public reporting on social responsibility performance is something that the communicator should do. This demonstrates the organization's dedication to considering the interest of all stakeholders.

6.3.3 Facilitate Stakeholder Dialogue

Communication should be a two-way street. Professional communicators are in a great position to ensure that stakeholder dialogue doesn't become stakeholder monologue. Communicators can facilitate the listening part of stakeholder dialogue and deliver difficult messages and manage difficult speaking parts of stakeholder dialogue. They fulfill an important role of ensuring that dialogue is effective.

For example, let's say that a community focus group is held to discuss a factory expansion. Positive points would include increased employment and employment security. But negative points might include increased noise, traffic, and energy consumption. Ideally, this stakeholder engagement event is facilitated in a way that provides balance for both the positive and negative aspects of the expansion. The communicator as facilitator can ensure that this event does not devolve into a complaint session about the increased traffic, nor becomes a blind celebration on job creation. The communicator can act to facilitate collaborative solutions to problems where practical, or considerations of changes where some negative impact in necessary as a part of a larger positive impact.

6.3.4 Stop the Spin

In the very early days of social responsibility, back when activities were referred to as CSR, or even corporate philanthropy, sometimes the motivation for such activity was reputation building. The organization engaged in philanthropy as a part of its public relations strategy. Luckily, we've moved on from this approach. Now we see social responsibility as a way to ensure sustainable business success.

Unfortunately, some organizations have not moved on from the public relations "spin." This becomes obvious to even casual stakeholders when the organizations community outreach or philanthropy seems to have no connection to its business strategy. This is activity completed under the guise of social responsibility, but no true responsibility to stakeholders is present. This activity is also often typified by large-scale advertisements and press releases, often oversized in comparison to the communities or stakeholder involved.

Some examples include an oil company, which has previously been found responsible for devastating wet lands due to oil spills, advertising about how many people it is employing to clean up the wetlands. This type of press often backfires. The community is wise to the fact that the cleanup would not have been needed if oil spills had been responsibly prevented. Another example would be a fast food restaurant receiving negative reviews in social media because of an inorganic ingredient in their bread products. We find the organization suddenly advertising that they have a new, improved, healthier bread recipe, without reference to correcting their original irresponsibility. Our advice for the communicators in the organization is to avoid using communication to cover up irresponsibility or worse, as in these examples, try to attempt to appear responsible to the unsuspecting stakeholder *after* it has behaved irresponsibly. This type of communication is a cover up, or "spin."

6.3.5 Look for Active and Passive Stakeholder Information

Organizations engage with stakeholders in many ways. Some engagement is active, such as focus groups, customer product returns, town meetings, or employee participation. Other engagement is passive such as community activist organizations, or labor representation group activity. Information about stakeholder attitudes, interests, concerns, and activities is an important output of these types of engagement. Passive stakeholder information comes to the organization whether asked for or not. Active stakeholder information needs to be carefully nurtured and sought after. Both types of engagement and both types of information are important.

Let's use the stakeholder group of employees as an example. Active information from employees to the organization about attitudes, needs, or interests may come through the interaction of employees on improvement projects, through employee suggestion or feedback programs, or through targeted focus groups or meetings with employees held for the express purpose of seeking stakeholder information. Passive information from employee stakeholders may be found in sources such as attrition and retention data, labor representation activities and membership, and employee development trends and training status.

The more future-looking stakeholder information is, the better it can be used to understand the risk of irresponsibility to that stakeholder group. As with the employee stakeholder example, future-looking, and passive, stakeholder information might include secondary school areas of study or grades achieved; these are future employees and understanding the interests of these future stakeholders may better prepare engagement with these people once they become potential employees. Another example of future-looking, passive stakeholder information on employees might be understanding community employment levels. Employees who have spouses that have gained or lost a job, can significantly impact the attitudes of the employee. The communicator can have an important role by bringing active, passive, and future-looking stakeholder information to the organization.

6.3.6 *Connect People and Initiatives within the Organization*

Communicators are often great connectors. They often have large networks of acquaintances; they engage with a broad swath of stakeholders. And many of these stakeholders are inside the organization. The stakeholders may be employees, owners or shareholders, contractors or vendors who work within the organization on a regular basis, or close service groups such as on-site healthcare providers or cafeteria workers. Because of the stakeholder's day-to-day busyness they often don't get to interact with each other. The employee rushes to get lunch and get back to her desk, not realizing that the cafeteria server has many of the same interests in the sustainability of the organization that she does. The truck driver loads and unloads quickly to get back on the road for the next delivery, not realizing the final product quality inspector releasing product for shipment shares many interests with the truck driver. The communicator can help connect people, help recognize common interests, and provide information to build social responsibility initiatives that address the priorities of multiple stakeholder groups. Because of their networking skills, connecting people and initiatives in the organization is something the communicator can do today as social responsibility action.

6.3.7 Balance Business Needs with Reporting

We've said that the organization should report social responsibility performance publicly. This demonstrates transparency and accountability to responsible behavior. We've also said stop the "spin." Don't communicate just to attempt to appeal to stakeholder interests in an attempt to cover up irresponsible behavior. Now, we'll add a third complexity to the communicator's job. Only report what advances business needs. Ensure that external communications achieve a business strategy.

To achieve this careful balance, keep in mind the reason for social responsibility—sustainability. The business desires to be profitable and productive for over 1500 years. Reporting should work toward the achievement of this ideal. There are many strategic business reasons for reporting. One example of the need to report business strategy through social responsibility performance is to make sure that suppliers understand the long-term goals of the organization, such as growth or expansion, so that the supplier can sustainably continue to supply raw materials. Another example of a business need served by reporting social responsibility performance is to ensure that the community, including public schools, is adequately preparing future employees. Reporting can be difficult. It is important because it demonstrates stakeholder engagement, transparency, and accountability; and it can backfire if it comes across as insincere or inaccurate. Ensuring that there is a business need served by the reporting helps the communicator report that which helps the organization achieve sustainability.

6.3.8 Connect Any Philanthropy to Business Strategy

Social responsibility isn't synonymous with corporate philanthropy, in fact if applied incorrectly corporate philanthropy can be irresponsible. However, we are not saying that corporate philanthropy should be avoided. We are saying that any corporate philanthropy should be crafted to facilitate business strategy to achieve sustainable profitability. For example, a manufacturing organization that donates funds to the city ballet company has a very weak connection between philanthropy and sustaining the manufacturing business. It could be argued that having the ballet attracts top talent to the city, and therefore the manufacturer; but it's a weak argument. This type of philanthropy is best left to either the individual citizens of the city, or the ballet shoe manufacturing company. A better target for the manufacturing company's philanthropy might be the local trade college. The organization might donate money to build new laboratories or shops for a trade college. This example of philanthropy can be strongly linked to the sustainability of the success of the manufacturing company. Often organizational communicators are in roles closely connected to corporate affairs and corporate philanthropy. Communicators can play an important role by ensuring that philanthropic activities can be strongly linked to business success.

6.3.9 Be a Participant in Projects

The SOFAIR method is a team-oriented problem solving and improvement method. We have found that the best teams are diverse teams, teams that have members from diverse backgrounds and ways of thinking. These teams often have members unfamiliar with the process being studied; they have "fresh eyes" and they question the status quo. The communicator may be a fresh eyes team member. One thing the communicator can do today as social responsibility action is to be on a SOFAIR team.

6.3.10 Communicate Praises of Social Responsibility Improvers in the Field

SOFAIR projects are not easy. They take many months of concerted effort by many people. Stakeholder engagement takes a lot of time, organization, and energy. The analytic methods take a lot of training. Project outcomes cause change in the organization; change is usually difficult and disrupting. Project team members achieve heroic performance improvement through enormous effort. They could use a cheerleader. The communicator can be their cheerleader.

As an example, the communicators can organize project closure celebrations. They can recognize team members for training achievements through certification ceremonies. They can broadcast project success through newsletters. They can ensure that SOFAIR project leaders have opportunities to present their methods and outcomes with executives and top leadership. Communicators can be a key organizational member to cause the continued acceleration of CISR efforts by motivating the activities of team leaders and members through rewards and recognition of the efforts of the social responsibility improvers, working hard in the field.

6.4 Ten Things a Team Member Can Do Today as Social Responsibility Action

The CISR methodology and the SOFAIR method are team-oriented approaches. Cross-functional diverse teams can create impressive innovation. And participating on a team can be fun, thus, motivating continued participation. SOFAIR project team members have an important role towards achieving social responsibility performance improvement. Team member groups are large groups. They can have a significant influence on the organizational culture change. In addition to team participation, team members can contribute in many other ways as well. Here are 10 things a team member can do today as social responsibility action.

6.4.1 Participate in Social Responsibility Improvement

Participating on a SOFAIR project team can be fun and rewarding. These teams should be full of cross-functional organizational members from varied backgrounds and diverse approaches. The interaction and activity on a SOFAIR project team can be tough, rewarding work. And knowing that you are collaborating with others to improve social responsibility performance is satisfying. Volunteer to be on a project team. Volunteer to be on several project teams, and once one project is complete sign up for the next one.

The ultimate CISR goal is to embed a culture that results in achieving a sustainably profitable organization. To achieve this goal the organizational culture should be one permeated with sustainable decisions and actions. To achieve this culture, everyone in the organization must think sustainably. The goal of CISR is not to do SOFAIR project forever, but to use SOFAIR projects as a way to engage and train organizational members to think sustainably. So, by participating on SOFAIR projects, the team member is learning how to think sustainably. This on-the-project training will transfer to the other responsibilities of the team member's job. And little by little over time the organizational culture shifts. Participating in social responsibility improvement causes the organizational culture shift needed to achieve sustainability.

6.4.2 *Recognize Social Responsibility Risk in Your Surroundings*

On-the-project training has another advantage for the team members. After participating in many stakeholder engagement events, completing stakeholder analyses, and failure mode effects and analysis we become more aware of social responsibility threats in our surroundings. We begin to see risk in our day-to-day job. We begin to see risk in what those around us are doing. Becoming better at recognizing social responsibility risk in your surroundings is a great way for team members to contribute.

For example, let's say you are an accountant in a mid-sized hospital. And you've participated on three different SOFAIR projects as a team member. The first one tackled the subject of human rights and the issue of discrimination. The team's goal was to prevent discrimination in hiring practices, specifically amongst nurse aids with immigrant backgrounds. The second project concerned labor practices and the issue of work conditions. The team wanted to address potential safety concerns about the hospital's 12-hour shift schedules. And the third team with which you participated looked at improving the consumer issue of privacy. You worked to help improve database safety of patient records.

After these rich and important opportunities you have a better appreciation for the principles and subjects of social responsibility. You now know to ask questions about an off-shore equipment supplier's anticorruption practices. You now print fewer draft reports in consideration of saving paper. You now recognize the importance of internal auditing and system testing for your accounting systems. And you are now aware of the need to mentor students at your local community college in basic accounting practices and personal budgeting and financing skills. Your awareness of risk has increased through the application of social responsibility improvement.

6.4.3 Report Social Responsibility Threats

It's not enough to just be aware of social responsibility risk. When found we must take action to do something about threats. The mitigation of risks may or may not fall in the realm of responsibility of an individual team member. Effective methods of reporting social responsibility threats are needed and are needed to be used. Many organizations have an anonymous hotline phone number to which employees can report concerns. This can be one solution to reporting threats. Other organizations might have a suggestion box. And others may have an open-door policy for reporting concerns. Reporting risk should be an easy process with no negative repercussions to those doing the reporting.

It is also important for the team member to know that just because he has recognized a threat and reported it, does not mean there will be immediate action to resolve it. Mitigating the risk identified needs to be a part of the prioritization process, such as the FMEA. The magnitude of loss or the probability of occurrence may indicate that other risks need to be resolved first and no organization has the infinite resources needed to resolve all risk simultaneously. As the team member, report the threat and then trust the prioritization process to mitigate the risk at the appropriate time.

6.4.4 Take a Systems Perspective

As a part of becoming more aware of the social responsibility risk around you, understanding that all of our activities are connected as parts of systems is an important growth in understanding. Classical systems theory teaches us that for any transformative activity there is an input, a transform, an output and a feedback loop. Start to recognize these systems in your surroundings. Your awareness of the interconnectedness of activity will increase.

For example, recognize desk work as a part of the larger system. Think about the inputs to desk work: paper, pen, computer, phone, electricity, lights, chairs, buildings, roads, water, sewer, and on and on. Think about the outputs: information, reports, feedback, communication, transactions, waste paper, stored electronic files, computer storage systems, downstream jobs, profits for the business, etc. Think about the feedback loops: customer complaints, calls for reports or results, and a full in-box. Recognize that all you do is embedded in a system and that every part of this system has an impact on sustainability. Taking a systems perspective will raise your awareness toward opportunities to improve social responsibility performance.

6.4.5 Learn the Tools

We have introduced many different analytic and problem solving tools in this book as a part of the SOFAIR method. All of these tools can be used independently without the formality of a SOFAIR project. For example, us an FMEA on your decisions about purchasing cleaning supplies for your home. Use the SIPOS to better understand the process and stakeholders at the next football team fundraising bake sale for your child's school. Use a prioritization matrix for your new home purchase. Use 5-whys to better understand neighborhood security. The more practiced you are at using the SOFAIR tools, the easier they will be to apply toward social responsibility performance improvement. Learn the tools and practice them frequently in all sorts of settings.

6.4.6 Practice Social Responsibility at Work and Home

All sorts of settings lead us to the need to practice social responsibility at work and home. Live an authentically socially responsible life. In harsher words, don't be a hypocrite. Don't work on a SOFAIR project at work to ensure legal employment status for cafeteria workers, and then hire someone with illegal status to mow your lawn. Don't lead a project to reduce electricity use in an industrial setting and then fail to adequately insulate your home. Don't work toward transparent organizational governance at work and then make unethical decision at home. Do practice all the steps of SOFAIR at home. Do analyze your responsibility risks at home. Do use the tools and techniques at home. Living authentically will breathe freedom and satisfaction into all that you do at work and home.

6.4.7 Get Active in Your Community

Community involvement and development is one of the core subjects of social responsibility. You are your community. Get involved; get active in what interests you in your community. This activity could be for fun, charity, or skill development. For example, joining the local municipal golf league can be a fun way to meet other people who golf, enjoy the game of golf, and better understand and help protect the local environment from unnecessary herbicides and fertilizers. You'll want to golf in a beautiful environment; joining the golf club will motivate you to be more interested in protecting the environment. Getting involved in charitable organizations will help you better understand the status of the most vulnerable groups in your community. For example, volunteering to organize a blood drive will help you understand the magnitude of those in healthcare crisis and in need of blood. Maybe you want to sharpen your skills in a topic unrelated to your job. Maybe you want to learn to write fiction; you've always dreamed of writing a novel. Seek out a creative writing class at your local community center. This will help you to recognize the need for community centers and diverse participants, both as givers and receivers, in community spaces. Getting active in your community for fun, charity, or skill development will expand your understanding of community stakeholders.

6.4.8 Leverage Team Leadership Skills

If you've had the privilege of being trained as a SOFAIR project leader, then you also have the responsibility to leverage your team leadership skills in other ways. Leading a SOFAIR team is not an easy task. You have people with disparate opinions, attitudes, and backgrounds on your team. You need to find known and unknown risks. You need to coalesce business, stakeholder, and environmental issues. You need to be hyper-creative to find solutions that improve performance for all stakeholders. The SOFAIR project leader quickly becomes an expert leader.

Leadership skills might be one of the most needed, yet least celebrated, set of skills. Leaders keep team members focused on the task at hand, without stifling participation. Leaders push for breakthrough innovation, without demanding the impossible. Leaders motivate engagement, while setting boundaries and realistic expectations. Effective leadership requires the skill to carefully assess situations, moment-by-moment, and apply pressure for performance only where needed. This is an astonishing skill. Once you have it, use it, relentlessly.

6.4.9 *Think about the Next Generation*

Whether you have children or not, concern for the next generation is a basic human tendency. If you are a parent you are intensely interested in the standard of living of your child's future. If we consider sustainability as a 1500 year view; then consider your children's, children's, children's, and so on, for at least 30 generations. What will your ancestors do in the year 3500 AD? Will they be living sustainably? They will only be living if the generations prior are living sustainably. Taking a 1500 year view and thinking about far future generations helps to influence our decisions away from instant gratification at great cost and toward sustainable decisions.

6.4.10 *Seek Perfection in the Day-to-Day*

Sometimes seeking sustainability can be overwhelming. There is risk everywhere we turn. There are compromises and currently unsustainable practices everywhere we look. We drive cars with combustion engines every day; someday the oil will run out; this is unsustainable. We purchase food in plastic containers; plastic does not biodegrade and the plastic trash can accumulate in the ocean or in landfills; this is unsustainable. There are many, many unsustainable practices in our modern lives. But, remember this is a journey. And a journey starts with one small step. As long as there is forward progress, little by little, over long periods of time, by many people, success will eventually happen. This is a marathon, not a sprint. Seek satisfaction in the day-to-day small decisions. One organic apple, one public transportation commute, one disagreement resolved, one law obeyed, one friend in need helped…this is the path to sustainability.

Appendix A: Glossary

Accountability: The recognition that no organization is perfect and a provision of trust in appropriate action when problems arise.

Analyze Phase of SOFAIR: The analyze phase is about thoroughly understanding all elements and interactions in the system that is the focus on improvement. Understanding causal factors before changes are made is critical for robust improvement. Part of the rigor of the SOFAIR method is the requirement to understand cause and effect before process change is enacted in order to ensure that performance improvement (not just process change) is implemented.

CISR® (Sounds like scissor): Abbreviation for Continual Improvement for Social Responsibility and registered trademark, which can be used with permission for non-commercial use. Contact SherpaBCorp.com for permission.

Community Involvement and Development: Defined as a core subject in ISO 26000, community involvement and development concerns the active engagement of the organization in the concerns of the community in which the organization operates and the responsibility to create improvements in the community.

Consumer Issues: Defined as a core subject in ISO 26000, consumer issues are the protective responsibilities to the consumers of the organization's goods or service.

Continual Improvement: A business management strategy that expects constant analysis and improvement through regimented change process and breakthrough innovation when needed.

Corporate Philanthropy: The practice of corporate cash donations to charitable organizations.

Corporate Social Responsibility (see Social Responsibility): The application of social responsibility to the subgroup of organizations chartered as corporations.

CSR (see Corporate Social Responsibility).

Critical-to-Sustainability Characteristic (CTS): A measurable characteristic that takes into consideration stakeholder wants and needs.

Environment: As defined as a core subject in ISO 26000, the environment considers the impact on the natural environment arising from the decisions and activities of the organization.

External Context: The business environment and conditions, external to the organization, that can have an impact on the decisions and actions of the organization.

Ethical Behavior: Acting with integrity, honesty, fairness, and the concern for all stakeholders and the environment.

Failure Mode, Effects, and Analysis: A specific type of risk assessment where failure modes are identified. The effects of these failure modes are also identified. The failure modes are ranked by their relative severity and occurrence.

Fair Operating Practices: Defined as a core subject in ISO 26000, fair operating practices are those dealing with other organizations completed in an ethical manner.

Fishbone Diagram: An organizational technique used to manage potential causal factors during root cause analysis brainstorming.

Focus: A subphase in the SOFAIR method where it is determined which process steps will be studied and which process steps will not be of focus. The focus element of the SOFAIR method recognizes that for any given project only a small part of the system will be the focus for improvement. The focus of SOFAIR allows us to tackle manageable size opportunities for social responsibility performance improvement.

FMEA (see Failure Mode, Effects, and Analysis).

Function: A subphase in the SOFAIR method, where tools and techniques to prioritize and select the most important part of the organization in which to focus the project, is determined. Evaluating the functions within an organization or process begins the adoption of systems thinking. Understanding the functions within the organization allows us to apply systems thinking.

Hoshin Kanri: A method of aligning strategic performance improvement through all levels of an organization. Also known as strategy deployment.

Human Rights: As defined as a core subject in ISO 26000, human rights are the basic rights to which all human beings are entitled because they are human beings.

Impacts: The influence, effect, or outcome of an action. With respect to social responsibility, these are the effects of behaviors on the social and ecological systems in the surrounding environment.

Improve: A subphase of the SOFAIR method when process change takes place. The end result of the Improve phase is an action plan, implemented, that yields more socially responsible performance. Effective project management and communication is the purpose of the Improve phase. The Improve phase focuses on managing and controlling the process changes as they happen.

Innovate: A subphase of the SOFAIR method used when an entirely new way of approaching the process is needed. The process must be intentionally broken and reset. The desired social responsibility simply can't be achieved with the current process.

Internal Context: The business environment and conditions, internal to the organization, that can have an impact on the decisions and actions of the organization.

Kaizen: A term used in relation to the Toyota Production System to mean small, incremental, but continual process improvements to be completed by everyone in the organization.

Labor Practices: As defined as a core subject in ISO 26000, labor practices encompass all policies and practices relating to work performed within, by or on behalf of the organization.

Lean Production: A colloquial term used to indicate the business management strategy deployed by Toyota Motor Corp. This involves a focus on the elimination of nonvalue added activity and waste from processes.

Lifecycle: A consideration of the full range of a product's or service's impact from raw material to after-use disposal or reuse.

Materiality: A term used in the investment community, and borrowed by the Global Reporting Initiative. Originally meaning that a fact is material if there is a substantial likelihood that the fact would have been viewed by the reasonable investor as having significantly altered the "total mix" of information made available. In GRI terms, it means that a topic has meaning and interest by a stakeholder.

Narrative: A written description. In reference to social responsibility, it is a description of the boundaries, assumptions, goals, and intentions toward a target of improvement.

Non-Governmental Organization (NGO): A social organization of treaty, guideline, or behavioral importance, but not related to any specific government or sovereignty.

Objective: A subphase in the SOFAIR method. Objective refers to the organization's objective for embarking on social responsibility performance improvement. This objective is connected to business strategy. The SOFAIR method sets the expectation that very early on in the performance improvement process, the organization formalizes and documents the objective outcome that it expects to achieve by conducting the continual improvement work.

Organizational Governance : Defined as a core subject in ISO 26000, organizational governance is the system by which an organization makes and implements decisions.

Production Planning Process (3P): An innovation methodology used to introduce creative design into lean production methods.

Plan–Do–Study–Act: An iterative process of problem solving made popular by W. Edwards Deming. The problem solver plans to act upon the system in order to correct a problem, does the corrective action, checks for the outcome of the correction, and then acts upon the outcome to either plan additional correction or control the appropriate correction.

Process : An assembly of many actions in series and parallel, typified by inputs, outputs, and transforming activities.

Profound Knowledge: W. Edward Deming's concept of knowledge needed by managers of business and operational processes. It includes the appreciation of a system, knowledge of variation, theory of knowledge, and knowledge of psychology. It is only after the achievement of profound knowledge that a process should become the target of continuous improvement.

Quality Function Deployment (QFD): A method translating customer needs for a product's fit, form, or function to manufacturing or design features.

Repeat: A subphase in the SOFAIR method. This is a way of introducing continual improvement in the method. When one project closes, the next project is opened. Untangling small pieces of problems with discrete but continuing effort can make solving complex and difficult problems possible.

Report: A subphase of the SOFAIR method. Reporting can be internal, external, or both. Formal sustainability reporting can be separate from other business and financial reporting, or integrated. And there are many reporting frameworks that can be adopted. Reporting helps to communicate and instill accountability for the sustainment and continual improvement of social responsibility performance improvement.

Responsibility: Responsibility refers to an ownership or accountability assumed; responsibility is voluntary accountability.

Risk: An unknown, positive or negative, future state.

Root Cause Problem Solving: The process of uncovering the fundamental conditions and actions that leads to problematic outcomes.

SIPOS (see Suppliers–Inputs–Process–Outputs–Stakeholders).

Six Sigma: A business management strategy that targets the reduction of process variation and the achievement of the elimination of defects.

Social Responsibility (SR): The responsibility of an organization for the impacts of its decisions and activities on society and the environment, through transparent and ethical behavior that contributes to sustainable development, health, and the welfare of society; which takes into account the expectations of stakeholders, is in compliance with applicable law and consistent with international norms of behavior, and is integrated throughout the organization and practiced in its relationships. This includes products, services, and processes. Relationships refer to an organization's activities within its sphere of influence (from ISO 26000).

SR (see Social Responsibility).

Stakeholder: A person or group which can be affected by the organization's activities and decisions.

Subjects: A subphase in the SOFAIR method. The subjects are defined by ISO 26000. There are seven subjects: organizational governance, human rights, labor practices, environment, fair operating practices, consumer issues, and community involvement and development.

Sustainability: The ability of social and ecological systems to endure over very long periods of time. Sustainable systems do not exhaust their required inputs without the reuse or recycle of system output back to input.

Suppliers–Inputs–Process–Outputs–Stakeholders (SIPOS): A graphical technique of listing all the suppliers, inputs, outputs, and stakeholders of a process. It also includes a summarized, graphical depiction of the process.

The Quality Movement: The period of time, from post-World War II to the 1980s, in which business management strategy moved from arts and crafts and command and inspection, to process control, statistical analysis, and process-oriented problem solving. It was this period of time when significant increases in manufactured product quality were achieved.

Transparency: The act of conducting business decision in a nonsecret environment. Transparency involves open access by stakeholders to the information and process involved in organizational decision making.

Appendix B: Additional References for CISR Practitioners

Automotive Industry Action Group. 2008. *Potential Failure Mode and Effects Analysis: (FMEA)* (4th Ed.). Detroit, MI: Automotive Industry Action Group. ISBN: 9781605341361.

Aven, T. 2003. *Foundations in Risk Analysis: A Knowledge and Decision-Oriented Perspective*. West Sussex, England: John Wiley & Sons, Ltd.

Bowler, C. and Brimblecombe, P. 2000. Control of air pollution in Manchester prior to the Public Health Act, 1875. *Environment and History*, 6, 71–98.

Brimblecombe, P. 1987. *The Big Smoke: A History of Air Pollution in London Since Medieval Times*. London: Methuen.

Cain, L. P. 1977. An economic history of urban location and sanitation. *Research in Economic History*, 2, 337–389.

Coles, G., Fuller, B., Nordquist, K., and Kongslie, A. 2005. Using failure modes analysis and criticality analysis for high-risk processes at three community hospital. *Joint Commission on Accreditation of Healthcare Organizations*, 31(3), 132–140.

Feigenbaum, A. V. 1961. *Total Quality Control*. New York, NY: McGraw-Hill.

Friedman, M. 1970. The social responsibility of business is to increase profits. *The New York Times Magazine*, September 13.

Frynas, J. G. and Pegg, S. 2003. *Transnational Corporations and Human Rights*. Houndsmills, Basingstoke, Hampshire RG21 6XS: Palgrave Macmillan.

Ford, R. 2005. Stakeholder leadership: Organizational change and power. *Leadership & Organization Development Journal*, 26, 616–638.

Gandy, M. 1994. *Recycling and the Politics of Urban Waste*. London: Earthscan Publications.

Goldman, J. A. 1997. *Building New York's Sewers: Developing Mechanisms of Urban Management.* West Lafayette, Ind.: Purdue University Press.

Halliday, S. 2001. *The Great Stink of London. Sir Joseph Bazalgette and the Cleansing of the Victorian Metropolis.* Stroud: Sutton Publishing.

Handfield, R. B. and McCormack, K. (Eds.). 2008. *Supply Chain Risk Management: Minimizing Disruptions in Global Sourcing.* New York, NY: Auerbach Publications.

Hoy, S. 1995. *Chasing Dirt: The American Pursuit of Cleanliness.* New York: Oxford University Press.

Kidder, R. 1997. Disasters chronic and acute: Issues in the study of environmental pollution in urban Japan. In: Pradyumna, P. K. and Stapleton, K. E. (Eds.), *The Japanese City.* Lexington: University Press of Kentucky, pp. 156–175.

Mayard, Y. M. 2007. Consumers' and leaders' perspectives: Corporate social responsibility as a source of a firm's competitive advantage (Doctoral Dissertation, University of Phoenix, 2007). Retrieved December 29, 2008, from Dissertations and Theses: Full Text Database. (UMI No. 3286256).

Melosi, M. 1980. *Pollution & Reform in American Cities, 1870–1930.* Austin: University of Texas Press.

Melosi, M. V. 1981. *Garbage in the Cities: Refuse, Reform and the Environment, 1880–1980.* College Station: Texas A&M University Press.

Melosi, M. V. (Ed.). 2000. *The Sanitary City: Urban Infrastructure in America from Colonial Times to the Present.* Baltimore: Johns Hopkins University Press.

Newton, L. and Schmidt, D. P. 2004. *Wake Up Calls: Classic Cases in Business Ethics.* Toronto, Ontario: Thomson South-Western.

Olsson, G. 2001. The struggle for a cleaner urban environment: Water pollution in Malmö 1850–1911. *Ambio,* 30, August, 287–291.

TarrStaib, R. 2005. *Environmental Management and Decision Making for Business.* Houndmills Basingstoke, Hampshire RG21 6XS: Palgrave Macmillan.

Mosley, S. 2001. *The Chimney of the World: A History of Smoke Pollution in Victorian and Edwardian Manchester.* Cambridge: White Horse Press.

Waldman, D., Lituchy, T., Gopalakrishnan, M., Laframboise, K., Galperin, B., and Kaltsounakis, Z. 1998. A qualitative analysis of leadership and quality improvement. *Leadership Quarterly,* 9, 177–191.

Wasylyshyn, K. M. 2008. Behind the door: Keeping business leaders focused on how they lead. *Consulting Psychology Journal: Practice and Research,* 60, 314–330.

Wilson, R. and Crouch, E. A. C. 2001. *Risk–Benefit Analysis.* Boston, MA: Harvard University Press.

Appendix C: References

Akao, Y. 2004. *QFD: Quality Function Deployment—Integrating Customer Requirements into Product Design.* New York, NY: Productivity Press.

Altshuller, G. 1996. *And Suddenly the Inventor Appears: TRIZ, The Theory of Inventive Problem Solving.* Worcester, MA: Technical Innovations Center.

The American Society for Quality. n.d. *The Value of an ASQ Certification.* Retrieved from http://asq.org/cert/control/index?gclid=COeh_p-J_cYCFVFsfgod0ikO3w.

The American Society for Quality. n.d. *W. Edwards Deming: A Mission Pursued on Two Continents.* Retrieved from http://asq.org/about-asq/who-we-are/bio_deming.html.

Babich, P. 2005. *Hoshin Handbook*, 3rd ed. Poway, CA: Total Quality Engineering.

Bader, C. 2015. *What Do Chief Sustainability Officers Actually Do*, May 6. Retrieved from The Atlantic: http://www.theatlantic.com/business/archive/2015/05/what-do-chief-sustainability-officers-actually-do/392315/.

Brundtland, U. G. 1987. *Report of the World Commission on Environment and Development: Our Common Future.* Retrieved from http://www.un-documents.net/our-common-future.pdf.

Byrd, H. 2007. *A Comparison of Three Well Known Behavior Based Safety Programs: DuPont STOP Program, Safety Performance Solutions and Behavioral Science Technology.* Rochester, NY: Rochester Institute of Technology.

Coletta, A. R. 2012. *The Lean 3P Advantage: A Practitioner's Guide to the Production Preparation Process.* Boca Raton, FL: CRC Press.

Dettmer, H. W. 2007. *The Logical Thinking Process: A Systems Approach to Complex Problem.* Milwaukee, WI: American Society for Quality.

Duckworth, H. 2010. *Social Responsibility: A Phenomenology of Perceived-Successful Private Sector Leadership Experience.* Doctoral dissertation UMI No. 3423856.

Duckworth, H. A. and Moore, R. A. 2010. *Social Responsibility: Failure Mode Effects and Analysis.* Boca Raton, FL: CRC Press.

Deming, W. E. 2000. *The New Economics for Industry, Government, Education* 2nd ed. Boston, MA: The MIT Press.

Deming's Fourteen Points for Management. n.d. In *The W. Edwards Deming Institute.* Retrieved from https://www.deming.org/theman/theories/fourteenpoints.

Deming's PDSA Cycle. n.d. In *The W. Edwards Deming Institute.* Retrieved from https://www.deming.org/theman/theories/pdsacycle.

Elkington, J. 1998. *Cannibals with Forks: The Triple Bottom Line of 21st Century Business.* Stony Creek, CT: New Society Publishers.

The European Organization for Quality. 2015. *EOQ Professions.* Retrieved from http://www.eoq.org/uploads/media/EOQ_certificates_list_040715_web.pdf.

European Organization for Quality (EOQ). (n.d.). *EOQ Competence Specifications and Certification Schemes.* Retrieved from: http://www.eoq.org/eoq_competence_specifications_and_certification_schemes.html.

Gano, D. 1999. *Apollo Root Cause Analysis: A New Way of Thinking.* Kennewick, WA: Apollonian Publications.

Hutcheson, J. O. 2007. The end of a 1400-year-old business. *Bloomberg Business,* April 16. Retrieved from http://www.bloomberg.com/bw/stories/2007-04-16/the-end-of-a-1-400-year-old-businessbusinessweek-business-news-stock-market-and-financial-advice.

Favaro, K. 2015. A Brief History of the Ways Companies Compete. *The Harvard Business Review,* April 22. Retrieved at https://hbr.org/2015/04/a-brief-history-of-the-ways-companies-compete.

The Global Reporting Initiative. n.d. *G4 Sustainability Reporting Guidelines.* Retrieved from http://www.globalreporting.org/.

Goldratt, E.M. 2014. *The Goal: A Process of On-going Improvement,* 4th Ed. Great Barrington, MA: The North River Press Publishing Corporation.

Grenny, J. and Patterson, K. 2013. *Influencer: The New Science of Leading Change,* 2nd Ed. New York, NY: McGraw-Hill.

International Standards Organization. 2010. *Guidance on Social Responsibility.* ISO Publication ISO 26000:2010. Geneva, Switzerland: International Standards Organization.

The International Council of Toy Industries. n.d. *ICTI CARE Process Case Studies.* Retrieved from http://www.icti-care.org/e/content/cat_page.asp?cat_id=188.

The Investor Responsibility Research Institute. n.d. Retrieved from http://www.irrcinstitute.org/.

Juran, J. 1999. *The Juran Quality Handbook.* New York, NY: McGraw-Hill.

Katzenback, J. R., Steffen, I., and Kronley, C. 2012. Cultural change that sticks. *The Harvard Business Review,* July–August.

Landes, C. E. 2012. *The Risk Assessment Auditing Standards: How to Efficiently and Effectively Comply on Smaller and Less Complex Audit Engagements*. The American Institute of Certified Public Accountants. Retrieved from http://www.aicpa.org/InterestAreas/ FRC/AuditAttest/DownloadableDocuments/Risk_Assessment/ Risk_Assessment_WP.pdf.

Lovins, A. and Odum, M. 2011. *Reinventing Fire: Bold Business Solutions for the New Energy Era*. White River Junction, VT: Chelsea Green Publishing.

Martin, I. 2015. *Exploring Consumption Behavior in Sustainability Research*, March. Retrieved from American Marketing Association: https://www. ama.org/publications/E publications/Pages/ama-journal-reader-march-15-consumption-sustainability.aspx.

Military Standard. 1980. *Procedures for Performing a Failure Mode and Criticality Analysis, Mil-Std-1629A*. Washington, DC: United States Department of Defense.

O'Hara, W. T. 2004. *Centuries of Success: Lessons from the World's Most Enduring Family Businesses*. Avon, MA: Adams Media.

Project Management Institute. n.d. *Project Management Body of Knowledge*. Retrieved from http://www.pmi.org/PMBOK-Guide-and-Standards. aspx.

Roberts, D. 2014. Workers Continue to Strike at Nike and Adidas Supplier in Southern China. *Bloomberg Business*, April 17. Retrieved from http:// www.bloomberg.com/bw/articles/2014-04-17/workers-continue-to-strike-at-nike-and-adidas-supplier-in-southern-china.

Santos, F. 2013. Corporate Social Responsibility: The Key to Attracting & Retaining Top Talent. *Forbes*. Retrieved from http://www.forbes.com/ sites/insead/2013/11/12/corporate-social-responsibility-the-key-to-attracting-retaining-top-talent/.

Security and Exchange Commission. 1999. *Staff Accounting Bulletin: No. 99—Materiality*, 17 CFR Part 211. Retrieved from https://www.sec.gov/ interps/account/sab99.htm.

Shewhart, W. A. 1931. *Economic Control of Quality of Manufactured Product*. New York: D. Van Nostrand Company.

Social Accountability International. 2014. *Social Accountability 8000: International Standard*. Retrieved from http://sa-intl.org/_data/n_0001/ resources/live/SA8000%20Standard%202014.pdf.

Solid Creativity. n.d. *TRIZ40*. Retrieved from http://www.triz40.com/ TRIZ_GB.php.

TRIZ-Canada Organisation. 2015. *List of Companies Using TRIZ*. Retrieved from http://triz-canada.ca/home/index.php?option=com_content&view= article&id=53&Itemid=106.

Index